3/09

R. Smith

D1235887

Valentin Boju
Louis Funar

The Math Problems Notebook

Birkhäuser
Boston • Basel • Berlin

Valentin Boju
MontrealTech
Institut de Technologie de Montreal
P.O. Box 78575, Station Wilderton
Montreal, Quebec, H3S 2W9 Canada
valentinboju@montrealtech.org

Louis Funar
Institut Fourier BP 74
URF Mathematiques
Université de Grenoble I
38402 Saint Martin d'Heres cedex
France
funar@fourier.ujf-grenoble.fr

Cover design by Alex Gerasev.

Mathematics Subject Classification (2000): 00A07 (Primary); 05-01, 11-01, 26-01, 51-01, 52-01

Library of Congress Control Number: 2007929628

ISBN-13: 978-0-8176-4546-5   e-ISBN-13: 978-0-8176-4547-2

Printed on acid-free paper.

9 8 7 6 5 4 3 2 1

www.birkhauser.com   (TXQ/SB)

The Authors

Valentin Boju was professor of mathematics at the University of Craiova, Romania, until his retirement in 2000. His research work was primarily in the field of geometry. He further promoted biostatistics and biomathematics as a discipline within the Department of Medicine of the University of Craiova. He left Romania for Canada, where he has taught at MontrealTech since 2001. He was actively involved in coaching college students in problem-solving and intuitive mathematics. In 2004, he was honored with the title of Officer of the Order "Cultural Merit," Category "Scientific Research."

MontrealTech
Institut de Technologie de Montréal
P.O. Box 78575 Station Wilderton
Montréal, Quebec, H3S 2W9
Canada
e-mail: valentinboju@montrealtech.org

Louis Funar has been a researcher at CNRS, at the Fourier Institute, University of Grenoble, France, since 1994. During his college years, he participated three times in the International Mathematics Olympiads with the Romanian team in the years 1983–1985, winning bronze, gold, and silver medals. His main research interests are in geometric topology and mathematical physics.

Institut Fourier, BP 74
Mathématiques UMR 5582
Université de Grenoble I
38402 Saint-Martin-d'Hères cedex
France
e-mail: funar@fourier.ujf-grenoble.fr

To all schoolteachers, particularly to my wonderful parents, *Evdochia* and *Nicolae Boju*, who genuinely believed in the education of the younger generation.

To my parents, *Maria* and *Ioan Funar*, who laid the foundation for our collaboration, as a late reward for their hard work and help, for their enthusiasm, determinedness, and strength during all these years.

Preface

The authors met on a Sunday morning about 25 years ago in Room 113. One of us was a college student and the other was leading the Sunday Math Circle. This circle, within the math department of the University of Craiova, gathered college students who possessed a common passion for mathematics. Most of them were participating in the mathematical competitions in vogue at that time, namely the Olympiads. Others just wanted to have good time.

There were similar math circles everywhere in the country, in which high-school students were committed to active training for math competitions. The highest standard was achieved by the selective training camps organized at the national level, which were led by professional coaches and mathematicians who were able to stimulate and elicit high performance. To mention a few among the leaders, we recall Dorin Andrica, Toma Albu, Titu Andreescu, Mircea Becheanu, Ioan Cuculescu, Dorel Mihet, and Ioan Tomescu.

The Sunday Math Circle shifted partially from its purpose of Olympiad training by following freely the leader's ideas and thus becoming a place primarily intended for disseminating beautiful mathematics at an elementary level. The fundamental texts were the celebrated book of Richard Courant and Herbert Robbins with the mysterious title *What is mathematics?*, and the book of David Hilbert and Stefan Cohn–Vossen entitled *Geometry and the Imagination*. The participants soon realized the differences in both scale and novelty between the competition-type math problems that they encountered every day and the mathematics built up over hundreds of years and engaged in by professional mathematicians.

Competition problems have to be solved in a short amount of time, and people compete against each other. These might be highly nontrivial, unconventional problems requiring deep insights and a lot of imagination, but still they have the advantage that one knows in advance that they *have* a solution. In the real mathematical world, problems often had to wait not merely years but sometimes centuries to be solved. A working mathematician could go for months or years trying to solve a particular problem that had not been solved before and take the risk of bitterly failing. Moreover, the sustained effort needed for accomplishing such a task would have to take into account a delicate balance between the accumulation of knowledge, methods, and tools and the creation from scratch of a path leading to a solution. This intellectual adventure is filled with suspense and frustration, since one does not know for sure what one is trying to prove or whether it is indeed true.

Real mathematics seemed to many of us to be too far away from competition problems. The philosophy of the Sunday Math Circle was that, in contrast to what we might think, the border between the two kinds of mathematics could be so vague and flexible that at times one could cross it at an early age. The history of mathematics abounds in examples in which a fresh mind was able to find an unexpected solution that specialists had been unable to find. One needs, however, the right training and the unconventional sense of finding the inspiring problem. Eventually, once a solution has been found, mathematicians are then willing to try to understand it even better, and other solutions follow in time, each one simpler and clearer than the previous one. In some sense, once solved, even the hardest problems start losing slowly their aura of difficulty and eventually become just problems. Problems that are today part of the curriculum of the average high-school student were difficult research problems three hundred years ago, solved only by brilliant mathematicians. This phenomenon demonstrates the evolution that language and science have undergone since then.

The authors conceived the present book with the nostalgia of the "good old times" of the Sunday Math Circles. We wanted something that carries the mark of that philosophy, namely, a number of challenging math problems for Olympiads with a glimpse of related problems of interest for the mathematician. The present text is a collection of problems that we think will be useful in training students for mathematical competitions. On the other hand, we hope that it might fulfill our second goal, namely, that of awakening interest in advanced mathematics. Thus its audience might range from college students and teachers to advanced math students and mathematicians.

The problems in each section are in increasing order of difficulty, so that the reader give some of the problems a try. We wanted to have 25% easy problems concerning basic tools and methods and consisting mainly of instructional exercises. The beginner might jump directly to the solutions, where a concentrate of the general theory and basic tricks can be found. The largest chunk contains about 50% problems of medium difficulty, which could be useful in training for mathematical competitions from local to international levels. The remaining 25% might be considered difficult problems even for the experienced problem-solver. These problems are often accompanied by comments that put the results in a broader perspective and might incite the reader to pursue the research further.

The problems serve as an excuse for introducing all sorts of generalizations and closely related open problems, which are spread among the solutions. Some of these are truly outstanding problems that resisted the efforts of mathematicians over the centuries, such as the congruent numbers conjecture and the Riemann hypothesis, which are among seven Millennium Prize Problems that the Clay Mathematics Institute recorded as some of the most difficult issues with which mathematicians were struggling at the turn of the second millennium and offered a reward of one million dollars for a solution to each one. In mathematics the frontier of our knowledge is still open, and it abounds in important unsolved problems, many of which can be understood at the undergraduate level. The reader might be soon driven to the edge of that part of mathematics where he or she could undertake original research.

We drew inspiration from the spirit of the famous books by R. K. Guy and his collaborators. Nevertheless, our aim was not to build up a collection of open questions in elementary mathematics but rather to offer a journey among the basic methods in problem-solving. Developing intuition and strengthening the most popular techniques in mathematical competitions are equally part of the goal. In the meantime, we offer the enthusiastic reader a brief glimpse of mathematical research, a place where problems yet unsolved long for deliverance.

The present collection of problems evolved from a notebook in which the second-named author collected the most interesting and unconventional problems that he encountered during his training for mathematical competitions in the 1980s. In the tradition of the Romanian school of mathematical training, he encountered problems inspired by both Russian Olympiads and American Competitions. Some (if not most) of the problems have already appeared elsewhere, especially in *Kvant, Matematika v Shkole, American Mathematical Monthly, Elemente der Mathematik, Matematikai Lapok, Gazeta Matematica*, and so, in a sense, the collection gained a certain cosmopolitan flavor. We have given the source in the problem section, when known, and more detailed information in the comments within the solutions part.

The original set of problems is complemented by more basic exercises, which aim at introducing many of the tricks and methods of which the competitor should be aware. Years later, we followed the destiny of some of the most intriguing questions from the notebook. Some of them led to developments that are well beyond the scope of this book, and we decided to outline a few of them.

We have supplied a short glossary containing some less usual definitions and identities in the geometry of triangles and the solution of Pell's equation. In order to help the reader find his or her way through the book, we have provided both an index concerning all mathematical results needed in the proofs, which are usually stated at the place where they are used first, and an index of mathematical terms at the end of the book.

The authors have benefited from discussions, corrections, help, and feedback from several people, whom we want to thank warmly: Dorin Andrica, Barbu Berceanu, Roland Bacher, Maxime Wolff, Mugurel Barcău, Ioan Filip, and Simon György Szatmari. We also thank Ann Kostant, Editorial Director, Springer, and Avanti Paranjpye, Associate Editor, Birkhäuser Boston, for their productive suggestions. One of them led to the "Index of Topics and Methods," which might be useful for a better reception of the book by readers. We are very grateful to Elizabeth Loew, our Production Editor, for her patience and dedication to accuracy and excellence. Finally, we are thankful to David Kramer for his thorough copyediting corrections.

Valentin Boju and Louis Funar
Montreal, Canada and *Saint-Martin-d'Hères, France*
July 2006

Contents

Part I

Problems

1

Number Theory

Problem 1.1. Show that we have $C_n^k \equiv 0 \pmod 2$ for all k satisfying $1 \le k \le n - 1$ if and only if $n = 2^\beta$, where $\beta \in \mathbb{Z}_+^*$. Here C_n^k denotes the number of combinations, i.e., the number of ways of picking up a subset of k elements from a set of n elements. Known also as the binomial coefficient or choice number and sometimes denoted as $\binom{n}{k}$ it is given by the formula

$$C_n^k = \binom{n}{k} = \frac{n!}{k!(n-k)!}$$

where the factorial $n!$ represents

$$n! = 1 \times 2 \times 3 \times \cdots \times n.$$

Problem 1.2. Let $P = a_n x^n + \cdots + a_0$ be a polynomial with integer coefficients. Suppose that there exists a number p such that:

1. p does not divide a_n;
2. p divides a_i, for all $i \le n - 1$;
3. p^2 does not divide a_0.

Then P is an irreducible polynomial in $\mathbb{Z}[x]$.

Problem 1.3. Given $m_i, b_i \in Z_+, i \in \{1, 2, \ldots, n\}$, such that $\gcd(m_i, m_j) = 1$ for all $i \ne j$, there exist integers x satisfying $x \equiv b_i \pmod{m_i}$ for all i. This result is usually known as the Chinese remainder theorem.

Problem 1.4. If $\gcd(a, m) = 1$, then it is a classical result of Euler that we have the following congruence:

$$a^{\varphi(m)+1} \equiv 0 \pmod m,$$

where $\varphi(m)$ is the Euler totient function, which counts how many positive integers smaller than m are relatively prime to m. Prove that this equality holds precisely for those numbers a, m such that for any prime number p that divides a, if p^k divides m then p^k also divides a.

Problem 1.5. *(Kvant)* Let $p > 2$ be a prime number and $a_k \in \{0, 1, \ldots, p^2 - 1\}$ denote the value of k^p modulo p^2. Prove that

$$\sum_{k=1}^{p-1} a_k = \frac{p^3 - p^2}{2}.$$

Problem 1.6. If $k \in \mathbb{Z}_+$, then show that

$$\left[\sqrt{k^2 + 1} + \cdots + \sqrt{k^2 + 2k}\right] = 2k^2 + 2k,$$

where $[x]$ denotes the integer part, i.e., the largest integer smaller than x.

Problem 1.7. *(Amer. Math. Monthly)* If n is not a multiple of 5, then $P = x^{4n} + x^{3n} + x^{2n} + x^n + 1$ is divisible by $Q = x^4 + x^3 + x^2 + x + 1$.

Problem 1.8. *(Amer. Math. Monthly)* Let p be an odd prime and $k \in \mathbb{Z}_+$. Show that there exists a perfect square the last k digits of whose expansion in base p are 1.

Problem 1.9. Any natural number greater than 6 can be written as a sum of two numbers that are relatively prime.

Problem 1.10. *(International Math. Olympiad)* Prove that it is impossible to extract an infinite arithmetic progression from the sequence $S = \{1, 2^k, 3^k, \ldots, n^k, \ldots\}$, where $k \geq 2$.

Problem 1.11. *(Amer. Math. Monthly)* Prove that b^{a-j+1} divides $C_{b^a}^j$ if $a, b \geq 2, j \leq a + 1$.

Problem 1.12. Solve in integers the following equations: (1) $x^2 = y^2 + y^3$; (2) $x^2 + y^2 = z^2$.

Problem 1.13. Let $a, b, c, d \in \mathbb{Z}_+$ be such that at least one of a and c is not a perfect square and $\gcd(a, c) = 1$. Show that there exist infinitely many natural numbers n such that $an + b, cn + d$ are simultaneously perfect squares if one of the following conditions is satisfied:

1. b and d are perfect squares;
2. $a + b, c + d$ are perfect squares;
3. $a(d - 1) = c(b - 1)$.

Moreover, there do not exist such numbers if $a = 1, b = 0, c = 4k^2 - 1, d = 1$.

Problem 1.14. If $N = 2 + 2\sqrt{28n^2 + 1} \in \mathbb{Z}$ for a natural number n, then N is a perfect square.

Problem 1.15. Let $n \geq 5, 2 \leq b \leq n$. Prove that

$$\left[\frac{(n-1)!}{b}\right] \equiv 0 \pmod{b - 1}.$$

Problem 1.16. *(Amer. Math. Monthly)* Prove that for every natural number n, there exists a natural number k such that k appears in exactly n nontrivial Pythagorean triples.

Problem 1.17. *(Amer. Math. Monthly)* Let $n, q \in \mathbb{Z}_+$ be such that all prime divisors of q are greater than n. Show that

$$(q - 1)(q^2 - 1) \cdots (q^{n-1} - 1) \equiv 0 \pmod{n!}.$$

Problem 1.18. Every natural number $n \geq 6$ can be written as a sum of distinct primes.

Problem 1.19. Every number $n \geq 6$ can be written as a sum of three numbers that are pairwise relatively prime.

Problem 1.20. *(Romanian training camp, Sinaia 1984)* Find all pairs of integers (m, n) such that
$$C_m^n = 1984,$$
where $C_m^n = \frac{m!}{n!(m-n)!}$ denotes the usual binomial coefficient.

Problem 1.21. *(Romanian training camp, Sinaia 1984)* Find the set A consisting of natural numbers n that are divisible by all odd natural numbers a with $a^2 < n$.

Problem 1.22. Prove that $2^{1092} - 1$ is divisible by 1093^2.

Problem 1.23. *(Amer. Math. Monthly)* Let $n \geq 0$, $r > 1$, and $0 < a \leq r$ be three integers. Prove that the number n, when written in base r, has precisely

$$\sum_{k=1}^{\infty} \left[nr^{-k} + ar^{-1} \right] - \left[nr^{-k} \right]$$

digits that are greater than or equal to $r - a$.

Problem 1.24. Find a pair (a, b) of natural numbers satisfying the following properties:

1. $ab(a + b)$ is not divisible by 7.
2. $(a + b)^7 - a^7 - b^7$ is divisible by 7^7.

Problem 1.25. *(International Math. Olympiad 1984)* Let $0 < a < b < c < d$ be odd integers such that

1. $ad = bc$
2. $a + d = 2^k, b + c = 2^m$, for some integers k and m.

Prove that $a = 1$.

Problem 1.26. *(Amer. Math. Monthly)* Find those subsets $S \subset \mathbb{Z}_+$ such that all but finitely many sums of elements from S (possibly with repetitions) are composite numbers.

Problem 1.27. *(Amer. Math. Monthly)* Prove that for any natural $n > 1$, the number $2^n - 1$ does not divide $3^n - 1$.

Problem 1.28. *(Putnam Competition 1964)* Let u_n be the least common multiple of the first n terms of a strictly increasing sequence of positive integers. Prove that

$$\sum_{n=1}^{\infty} \frac{1}{u_n} \le 2.$$

Find a sequence for which equality holds above.

Problem 1.29. *(Amer. Math. Monthly)* Let $\varphi_n(m) = \varphi(\varphi_{n-1}(m))$, where $\varphi_1(m) = \varphi(m)$ is the Euler totient function, and set $\omega(m)$ the smallest number n such that $\varphi_n(m) = 1$. If $m < 2^\alpha$, then prove that $\omega(m) \le \alpha$.

Problem 1.30. Let f be a polynomial with integer coefficients and $N(f) = \text{card}\{k \in \mathbb{Z}; f(k) = \pm 1\}$. Prove that $N(f) \le 2 + \deg f$, where $\deg f$ denotes the degree of f.

Problem 1.31. Prove that every integer can be written as a sum of 5 perfect cubes.

Problem 1.32. If $n \in \mathbb{Z}$, then the binomial coefficient $C_{2n} = \frac{(2n)!}{(n!)^2}$ has an even number of divisors.

Problem 1.33. *(Amer. Math. Monthly)* Prove that every $n \in \mathbb{Z}_+$ can be written in precisely $k(n)$ different ways as a sum of consecutive integers, where $k(n)$ is the number of odd divisors of n greater than 1.

Problem 1.34. *(Amer. Math. Monthly)* Let $\pi_2(x)$ denote the number of twin primes p with $p \le x$. Recall that p is a twin prime if both p and $p + 2$ are prime. Show that

$$\pi_2(x) = 2 + \sum_{7 \le n \le x} \sin\left(\frac{\pi}{2}(n+2)\left[\frac{n!}{n+2}\right]\right) \sin\left(\frac{\pi}{2}n\left[\frac{(n-2)!}{n}\right]\right)$$

for $x > 7$.

Problem 1.35. *(Kvant)*

1. Find all solutions of the equation $3^{x+1} + 100 = 7^{x-1}$.
2. Find two solutions of the equation $3^x + 3^{x^2} = 2^x + 4^{x^2}$, and prove that there are no others.

Problem 1.36. *(Kvant)* Let $\sigma(n)$ denote the sum of the divisors of n. Prove that there exist infinitely many integers n such that $\sigma(n) > 2n$, or even stronger, such that $\sigma(n) > 3n$. Prove also that $\sigma(n) < n(1 + \log n)$.

Problem 1.37. *(American Competition)* Let a_i be natural numbers such that $\gcd(a_i, a_j) = 1$ and the a_i are not prime numbers. Show that

$$\frac{1}{a_1} + \cdots + \frac{1}{a_n} < 2.$$

Problem 1.38. Let $\sigma(n)$ denote the sum of divisors of n. Show that $\sigma(n) = 2^k$ if and only if n is a product of Mersenne primes, i.e., primes of the form $2^k - 1$, for $k \in \mathbb{Z}$.

Problem 1.39. *(Putnam Competition)* Find all integer solutions of the equation $|p^r - q^s| = 1$, where p, q are primes and $r, s \in \mathbb{Z} \setminus \{0, 1\}$.

Problem 1.40. Consider an arithmetic progression with ratio between 1 and 2000. Show that the progression does not contain more than 10 consecutive primes.

Problem 1.41. *(Amer. Math. Monthly)* Let $a_1 = 1, a_{n+1} = a_n + \left[\sqrt{a_n}\right]$. Show that a_n is a perfect square iff n is of the form $2^k + k - 2$.

Problem 1.42. Recall that $\varphi(n)$ denotes the Euler totient function (i.e., the number of natural numbers less than n and prime to n), and that $\sigma(n)$ is the sum of divisors of n. Show that n is prime iff $\varphi(n)$ divides $n - 1$ and $n + 1$ divides $\sigma(n)$.

Problem 1.43. *(Amer. Math. Monthly)* A number is called φ-subadditive if $\varphi(n) \leq \varphi(k) + \varphi(n - k)$ for all k such that $1 \leq k \leq n - 1$, and φ-superadditive if the reverse inequality holds. Prove that there are infinitely many φ-subadditive numbers and infinitely many φ-superadditive numbers.

Problem 1.44. *(Amer. Math. Monthly)* Find the positive integers N such that for all $n \geq N$, we have $\varphi(n) \leq \varphi(N)$.

Problem 1.45. A number n is perfect if $\sigma(n) = 2n$, where $\sigma(n)$ denotes the sum of all divisors of n. Prove that the even number n is perfect if and only if $n = 2^{p-1}(2^p - 1)$, where p is a prime number with the property that $2^p - 1$ is prime.

Problem 1.46. *(Elemente der Mathematik)* A number n is superperfect if $\sigma(\sigma(n)) = 2n$, where $\sigma(k)$ is the sum of all divisors of k. Prove that the even number n is superperfect if and only if $n = 2^r$, where r is an integer such that $2^{r+1} - 1$ is prime.

Problem 1.47. If a, b are rational numbers satisfying $\tan a\pi = b$, then $b \in \{-1, 0, 1\}$.

Problem 1.48. *(Amer. Math. Monthly)* Let $A_n = ru^n + sv^n, n \in \mathbb{Z}_+$, where r, s, u, v are integers, $u \neq \pm v$, and let P_n be the set of prime divisors of A_n. Then $P = \bigcup_{n=0}^{\infty} P_n$ is infinite.

Problem 1.49. Solve in natural numbers the equation

$$x^2 + y^2 + z^2 = 2xyz.$$

Problem 1.50. *(Amer. Math. Monthly)* Find the greatest common divisor of the following numbers: $C_{2n}^1, C_{2n}^3, C_{2n}^5, \ldots, C_{2n}^{2n-1}$.

Problem 1.51. *(Amer. Math. Monthly)* Prove that if $S_k = \sum_{i=1}^{n} k^{\gcd(i,n)}$, then $S_k \equiv 0 \pmod{n}$.

Problem 1.52. Prove that for any integer n such that $4 \le n \le 1000$, the equation

$$\frac{1}{x} + \frac{1}{y} + \frac{1}{z} = \frac{4}{n}$$

has solutions in natural numbers.

Problem 1.53. *(Nieuw Archief v. Wiskunde)* For a natural number n we set $S(n)$ for the set of integers that can be written in the form $1 + g + \cdots + g^{n-1}$ for some $g \in \mathbb{Z}_+$, $g \ge 2$.

1. Prove that $S(3) \cap S(4) = \emptyset$.
2. Find $S(3) \cap S(5)$.

Problem 1.54. If $k \ge 202$ and $n \ge 2k$, then prove that $C_n^k > n^{\pi(k)}$, where $\pi(k)$ denotes the number of prime numbers smaller than k.

Problem 1.55. *(Amer. Math. Monthly)* Let $S_m(n)$ be the sum of the inverses of the integers smaller than m and relatively prime to n. If $m > n \ge 2$, then show that $S_m(n)$ is not an integer.

Problem 1.56. Solve in integers the following equations:

1. $x^4 + y^4 = z^2$;
2. $y^3 + 4y = z^2$;
3. $x^4 = y^4 + z^2$.

Problem 1.57. Show that the equation

$$x^3 + y^3 = z^3 + w^3$$

has infinitely many integer solutions. Prove that 1 can be written in infinitely many ways as a sum of three cubes.

Problem 1.58. Prove that the equation $y^2 = Dx^4 + 1$ has no integer solution except $x = 0$ if $D \not\equiv 0, -1, 3, 8 \pmod{16}$ and there is no factorization $D = pq$, where $p > 1$ is odd, $\gcd(p, q) = 1$, and either $p \equiv \pm 1 \pmod{16}$, $p \equiv q \pm 1 \pmod{16}$, or $p \equiv 4q \pm 1 \pmod{16}$.

Problem 1.59. *(Amer. Math. Monthly)* Find all numbers that are simultaneously triangular, perfect squares, and pentagonal numbers.

Problem 1.60. *(Amer. Math. Monthly)* Find all inscribable integer-sided quadrilaterals whose areas equal their perimeters.

Problem 1.61. *(Amer. Math. Monthly)* Every rational number a can be written as a sum of the cubes of three rational numbers. Moreover, if $a > 0$, then the three cubes can be chosen to be positive.

Problem 1.62. 1. Every prime number of the form $4m + 1$ can be written as the sum of two perfect squares.

2. Every natural number is the sum of four perfect squares.

Problem 1.63. Every natural number can be written as a sum of at most 53 integers to the fourth power.

Problem 1.64. Let $G(k)$ denote the minimal integer n such that any positive integer can be written as the sum of n positive perfect kth powers. Prove that $G(k) \geq 2^k + \left[\frac{3^k}{2^k}\right] - 2$.

Problem 1.65. *(Amer. Math. Monthly)* Let $R(n, k)$ be the remainder when n is divided by k and

$$S(n, k) = \sum_{i=1}^{k} R(n, i).$$

1. Prove that $\lim_{n \to \infty} \frac{S(n, n)}{n^2} = 1 - \frac{\pi^2}{12}$.
2. Consider a sequence of natural numbers (a_k) growing to infinity and such that $\lim_{k \to \infty} \frac{a_k \log k}{k} = 0$. Prove that $\lim_{k \to \infty} \frac{S(ka_k, k)}{k^2} = \frac{1}{4}$.

Problem 1.66. *(Nieuw Archief v. Wiskunde)* Let $a_1 < a_2 < a_3 < \cdots < a_n < \cdots$ be a sequence of positive integers such that the series

$$\sum_{i=1}^{\infty} \frac{1}{a_i}$$

converges. Prove that for any i, there exist infinitely many sets of a_i consecutive integers that are not divisible by a_j for all $j > i$.

Problem 1.67. *(Amer. Math. Monthly)* Denote by C_n the claim that there exists a set of n consecutive integers such that no two of them are relatively prime. Prove that C_n is true for every n such that $17 \leq n \leq 10000$.

Problem 1.68. *(Schweitzer Competition)* Prove that a natural number has more divisors that can be written in the form $3k + 1$, for $k \in \mathbb{Z}$, than divisors of the form $3m - 1$, for $m \in \mathbb{Z}$.

Problem 1.69. *(Amer. Math. Monthly)* A number N is called deficient if $\sigma(N) < 2N$ and abundant if $\sigma(N) > 2N$.

1. Let k be fixed. Are there any sequences of k consecutive abundant numbers?
2. Show that there are infinitely many 5-tuples of consecutive deficient numbers.

Problem 1.70. *(Amer. Math. Monthly)* Does there exist a nonconstant polynomial $an^2 + bn + c$ with integer coefficients such that for any natural numbers m all its prime factors p_i are congruent to 3 modulo 4? Prove that for any nonconstant polynomial f with integer coefficients and any $m \in \mathbb{Z}$ there exist a prime number p and a natural number n such that p divides $f(n)$ and $p \equiv 1 \pmod{m}$.

2

Algebra and Combinatorics

2.1 Algebra

Problem 2.1. Set $S_{k,p} = \sum_{i=1}^{p-1} i^k$, for natural numbers p and k. If $p \geq 3$ is prime and $1 < k \leq p - 2$, show that

$$S_{k,p} \equiv 0 \pmod{p}.$$

Problem 2.2. Let $P = a_0 + \cdots + a_n x^n$ and $Q = b_0 + \cdots + b_m x^m$ be two polynomials with $m \leq n$. Then, $\deg \gcd(P, Q) \geq 1$ if and only if there exist two polynomials K and L, such that $\deg K \leq m - 1$, $\deg L \leq n - 1$, and $K \cdot P = L \cdot Q$. Prove that this is equivalent to the vanishing of the following $(n + m) \times (n + m)$ determinant:

$$\det \begin{pmatrix}
a_0 & a_1 & a_2 & \cdots & a_{m-1} & \cdots & a_{n-1} & a_n & 0 & 0 & \cdots & 0 \\
0 & a_0 & a_1 & \cdots & a_{m-2} & \cdots & a_{n-2} & a_{n-1} & a_n & 0 & \cdots & 0 \\
\vdots & \vdots & \vdots & & \vdots & & \vdots & \vdots & \vdots & \vdots & & \vdots \\
0 & 0 & 0 & \cdots & a_0 & \cdots & a_{n-m-1} & a_{n-m} & a_{n-m+1} & \cdots & & a_n \\
0 & 0 & 0 & \cdots & 0 & \cdots & a_{n-m-2} & a_{n-m-1} & a_{n-m} & \cdots & & a_{n-1} \\
\vdots & \vdots & \vdots & & \vdots & & \vdots & \vdots & \vdots & & & \vdots \\
0 & 0 & 0 & \cdots & 0 & 0 & a_0 & a_1 & a_2 & \cdots & & a_m \\
b_0 & b_1 & \cdots & b_{m-1} & b_m & 0 & 0 & 0 & 0 & \cdots & & 0 \\
0 & b_0 & \cdots & b_{m-2} & b_{m-1} & b_m & 0 & 0 & 0 & \cdots & & 0 \\
\vdots & \vdots & & \vdots & \vdots & \vdots & \vdots & \vdots & \vdots & & & \vdots \\
0 & 0 & \cdots & b_0 & b_1 & \cdots & & \cdots & & 0 & \cdots & 0
\end{pmatrix} = 0.$$

Problem 2.3. Prove that if $P, Q \in \mathbb{R}[x, y]$ are relatively prime polynomials, then the system of equations

$$P(x, y) = 0,$$
$$Q(x, y) = 0,$$

has only finitely many real solutions.

Problem 2.4. Let $a, b, c, d \in \mathbb{R}[x]$ be polynomials with real coefficients. Set

$$p = \int_1^x ac\, dt, \quad q = \int_1^x ad\, dt, \quad r = \int_1^x bc\, dt, \quad s = \int_1^x bd\, dt.$$

Prove that $ps - qr$ is divisible by $(x-1)^4$.

Problem 2.5. 1. Find the minimum number of elements that must be deleted from the set $\{1, \ldots, 2005\}$ such that the set of the remaining elements does not contain two elements together with their product.

2. Does there exist, for any k, an arithmetic progression with k terms in the infinite sequence

$$1, \frac{1}{2}, \ldots, \frac{1}{2005}, \ldots, \frac{1}{n}, \ldots?$$

Problem 2.6. *(Amer. Math. Monthly)* Consider a set S of n elements and $n+1$ subsets $M_1, \ldots, M_{n+1} \subset S$. Show that there exist $r, s \geq 1$ and disjoint sets of indices $\{i_1, \ldots, i_r\} \cap \{j_1, \ldots, j_s\} = \emptyset$ such that

$$\bigcup_{k=1}^{r} M_{i_k} = \bigcup_{k=1}^{s} M_{j_k}.$$

Problem 2.7. Let p be a prime number, and $A = \{a_1, \ldots, a_{p-1}\} \subset \mathbb{Z}_+^*$ a set of integers that are not divisible by p. Define the map $f : \mathcal{P}(A) \to \{0, 1, \ldots, p-1\}$ by

$$f(\{a_{i_1}, \ldots, a_{i_k}\}) = \sum_{p-1}^{k} a_{i_p} \pmod{p}, \text{ and } f(\emptyset) = 0.$$

Prove that f is surjective.

Problem 2.8. Consider the function $F_r = x^r \sin rA + y^r \sin rB + z^r \sin rC$, where $x, y, z \in \mathbb{R}$, $A + B + C = k\pi$, and $r \in \mathbb{Z}_+$. Prove that if $F_1(x_0, y_0, z_0) = F_2(x_0, y_0, z_0) = 0$, then $F_r(x_0, y_0, z_0) = 0$, for all $r \in \mathbb{Z}_+$.

Problem 2.9. Let $T(z) \in \mathbb{Z}[z]$ be a nonzero polynomial with the property that $|T(u_i)| \leq 1$ for all values u_i that are roots of $P(z) = z^n - 1$. Prove that either $T(z)$ is divisible by $P(z)$, or else there exists some $k \in \mathbb{Z}_+$, $k \leq n - 1$, such that $T(z) \pm z^k$ is divisible by $P(z)$. The same result holds when instead of $P(z)$, we consider $z^n + 1$.

Problem 2.10.   1. If the map $x \mapsto x^3$ from a group G to itself is an injective group homomorphism, then G is an abelian.
2. If the map $x \mapsto x^3$ from a group G to itself is a surjective group homomorphism, then G is an abelian.
3. Find an abelian group with the property that $x \mapsto x^4$ is an automorphism.
4. What can be said for exponents greater than 4?

Problem 2.11. *(Amer. Math. Monthly)* Let V be a vector space of dimension $n > 0$ over a field of characteristic $p \neq 0$ and let A be an affine map $A : V \to V$. Prove that there exist $u \in V$ and $0 \leq k \leq np$ such that $A^k u = u$.

Problem 2.12. *(Putnam Competition 1959)* Find the cubic equation the zeros of which are the cubes of the roots of the equation $x^3 + ax^2 + bx + c = 0$.

Problem 2.13. *(Putnam Competition 1956)* Assume that the polynomials $P, Q \in \mathbb{C}[x]$ have the same roots, possibly with different multiplicities. Suppose, moreover, that the same holds true for the pair $P + 1$ and $Q + 1$. Prove that $P = Q$.

Problem 2.14. *(Amer. Math. Monthly)* Determine $r \in \mathbb{Q}$, for which 1, $\cos 2\pi r$, $\sin 2\pi r$ are linearly dependent over $\mathbb{Q}$.

Problem 2.15. *(Amer. Math. Monthly)*

1. Prove that there exist $a, b, c \in \mathbb{Z}$, not all zero, such that $|a|, |b|, |c| < 10^6$ $\left| a + b\sqrt{2} + c\sqrt{3} \right| < 10^{-11}$.
2. Prove that if $0 \leq |a|, |b|, |c| < 10^6$, $a, b, c \in \mathbb{Z}$, and at least one of them is nonzero, then $\left| a + b\sqrt{2} + c\sqrt{3} \right| > 10^{-21}$.

Problem 2.16. *(Amer. Math. Monthly)* Prove that if $n > 2$, then we do not have any nontrivial solutions for the equation

$$x^n + y^n = z^n,$$

where x, y, z are rational functions. Solutions of the form $x = af$, $y = bf$, $z = cf$, where f is a rational function and a, b, c are complex numbers satisfying $a^n + b^n = c^n$, are called *trivial*.

Problem 2.17. *(Kvant)* A table is an $n \times k$ rectangular grid drawn on the torus, every box being assigned an element from $\mathbb{Z}/2\mathbb{Z}$. We define a transformation acting on tables as follows. We replace all elements of the grid simultaneously, each element being changed into the sum of the numbers previously assigned to its neighboring boxes. Prove that iterating this transformation sufficiently many times, we always obtain the trivial table filled with zeros, no matter what the initial table was, if and only if $n = 2^p$ and $k = 2^q$ for some integers p, q. In this case we say that the respective $n \times k$ grid is nilpotent.

2.2 Algebraic Combinatorics

Problem 2.18. Let us consider a four-digit number N whose digits are not all equal. We first arrange its digits in increasing order, then in decreasing order, and finally, we subtract the two obtained numbers. Let $T(N)$ denote the positive difference thus obtained. Show that after finitely many iterations of the transformation T, we obtain 6174.

Problem 2.19. Find an example of a sequence of natural numbers $1 \leq a_1 < a_2 < \cdots < a_n < a_{n+1} < \cdots$ with the property that every $m \in \mathbb{Z}_+$ can be uniquely written as $m = a_i - a_j$, for $i, j \in \mathbb{Z}_+$.

Problem 2.20. *(Amer. Math. Monthly)* Consider the set of $2n$ integers $\{\pm a_1, \pm a_2, \ldots, \pm a_n\}$ and $m < 2^n$. Show that we can choose a subset S such that

1. The two numbers $\pm a_i$ are not both in S;
2. The sum of all elements of S is divisible by m.

Problem 2.21. *(Proposed for the International Math. Olympiad)* Show that for every natural number n there exist prime numbers p and q such that n divides their difference.

Problem 2.22. An even number, $2n$, of knights arrive at King Arthur's court, each one of them having at most $n - 1$ enemies. Prove that Merlin the wizard can assign places for them at a round table in such a way that every knight is sitting only next to friends.

Problem 2.23. *(Putnam Competition)* Let $r, s \in \mathbb{Z}_+$. Find the number of 4-tuples of positive integers (a, b, c, d) that satisfy $3^r 7^s = \operatorname{lcm}(a, b, c) = \operatorname{lcm}(a, b, d) = \operatorname{lcm}(a, c, d) = \operatorname{lcm}(b, c, d)$.

Problem 2.24. *(Putnam Competition)*

1. Let $n \in \mathbb{Z}_+$ and p be a prime number. Denote by $N(n, p)$ the number of binomial coefficients C_n^s that are not divisible by p. Assume that n is written in base p as $n = n_0 + n_1 p + \cdots + n_m p^m$, where $0 \leq n_j < p$ for all $j \in \{0, 1, \ldots, m\}$. Prove that $N(n, p) = (n_0 + 1)(n_1 + 1) \cdots (n_j + 1)$.
2. Write k in base p as $k = k_0 + k_1 p + \cdots + k_j p^j$, with $0 \leq k_j \leq p - 1$, for all $j \in \{0, 1, \ldots, s\}$. Prove that

$$C_n^k \equiv C_{n_0}^{k_0} C_{n_1}^{k_1} \cdots C_{n_j}^{k_j} \pmod{p}.$$

Problem 2.25. *(Putnam Competition)* Define the sequence T_n by $T_1 = 2, T_{n+1} = T_n^2 - T_n + 1$, for $n \geq 1$. Prove that if $m \neq n$, then T_m and T_n are relatively prime, and further, that

$$\sum_{i=1}^{\infty} \frac{1}{T_i} = 1.$$

Problem 2.26. Let $\alpha, \beta > 0$ and consider the sequences

$$[\alpha], [2\alpha], \ldots, [k\alpha], \ldots; [\beta], [2\beta], \ldots, [k\beta], \ldots,$$

where the brackets denote the integer part. Prove that these two sequences taken together enumerate $\mathbb{Z}_+$ in an injective manner if and only if

$$\alpha, \beta \in \mathbb{R} \setminus \mathbb{Q} \text{ and } \frac{1}{\alpha} + \frac{1}{\beta} = 1.$$

Problem 2.27. We say that the sets $S_1, S_2, \ldots, S_m$ form a complementary system if they make a partition of $\mathbb{Z}_+$, i.e., every positive integer belongs to a unique set S_i. Let $m > 1$ and $\alpha_1, \ldots, \alpha_m \in \mathbb{R}_+$. Then the sets

$$S_i = \{[n\alpha_i], \text{ where } n \in \mathbb{Z}_+\}$$

form a complementary system only if

$$m = 2, \quad \alpha_1^{-1} + \alpha_2^{-1} = 1, \quad \text{and } \alpha_1 \in \mathbb{R} \setminus \mathbb{Q}.$$

Problem 2.28. Let $f : \mathbb{Z}_+ \to \mathbb{Z}_+$ be an increasing function and set

$$F(n) = f(n) + n, \quad G(n) = f^*(n) + n,$$

where $f^*(n) = \text{card}(\{x \in \mathbb{Z}_+; 0 \le f(x) < n\})$. Then $\{F(n); n \in \mathbb{Z}_+\}$ and $\{G(n); n \in \mathbb{Z}_+\}$ are complementary sequences. Conversely, any two complementary sequences can be obtained this way using some nondecreasing function f.

Problem 2.29. Let M denote the set of bijective functions $f : \mathbb{Z}_+ \to \mathbb{Z}_+$. Prove that there is no bijective function between M and $\mathbb{Z}$.

Problem 2.30. *(Amer. Math. Monthly)* Let $F \subset \mathbb{Z}$ be a finite set of integers satisfying the following properties:

1. For any $x \in F$ there exist $y, z \in F$ such that $x = y + z$.
2. There exists n such that for any natural number $1 \le k \le n$, and any choice of $x_1, \ldots, x_k \in F$, their sum $x_1 + \cdots + x_k$ is nonzero.

Prove that $\text{card}(F) \ge 2n + 2$.

Problem 2.31. *(Amer. Math. Monthly)* For a finite graph G we denote by $Z(G)$ the minimal number of colors needed to color all its vertices such that adjacent vertices have different colors. This is also called the chromatic number of G.

Prove that the inequality

$$Z(G) \ge \frac{p^2}{p^2 - 2q}$$

holds if G has p vertices and q edges.

Problem 2.32. Let D_k be a collection of subsets of the set $\{1, \ldots, n\}$ with the property that whenever $A \ne B \in D_k$, then $\text{card}(A \cap B) \le k$, where $0 \le k \le n - 1$. Prove that

$$\text{card}(D_k) \le C_n^0 + C_n^1 + C_n^2 + \cdots + C_n^{k+1}.$$

Problem 2.33. Prove that

$$\frac{1}{p!} \sum_{k=0}^n (-1)^{n-k} C_n^k k^p = \begin{cases} 0, & \text{if } 0 \le p < n, \\ 1, & \text{if } p = n, \\ n/2, & \text{if } p = n+1, \\ \frac{n(3n+1)}{24}, & \text{if } p = n+2, \\ \frac{n^2(n+1)}{48}, & \text{if } p = n+3, \\ \frac{n(15n^3+30n^2+5n+1)}{1152}, & \text{if } p = n+4. \end{cases}$$

Problem 2.34. Write $\mathrm{lcm}(a_1, \ldots, a_n)$ in terms of the various $\gcd(a_i, \ldots, a_j)$ for subsets of $\{a_1, \ldots, a_n\}$.

Problem 2.35. Let $f(n)$ be the number of ways in which a convex polygon with $n+1$ sides can be divided into regions delimited by several diagonals that do not intersect (except possibly at their endpoints). We consider as distinct the dissection in which we first cut the diagonal a and next the diagonal b from the dissection in which we first cut the diagonal b and next the diagonal a. It is easy to compute the first values of $f(n)$, as follows: $f(1) = 1$, $f(2) = 1$, $f(3) = 3$, $f(4) = 11$, $f(5) = 45$. Find the generating function $F(x) = \sum f(n)x^n$ and an asymptotic formula for $f(n)$.

Problem 2.36. Find the permutation $\sigma : (1, \ldots, n) \to (1, \ldots, n)$ such that

$$S(\sigma) = \sum_{i=1}^{n} |\sigma(i) - i|$$

is maximal.

Problem 2.37. *(Amer. Math. Monthly)* On the set S_n of permutations of $\{1, \ldots, n\}$ we define an invariant distance function by means of the formula

$$d(\sigma, \tau) = \sum_{i=1}^{n} |\sigma(i) - \tau(i)|.$$

What are the values that d could possibly take?

Problem 2.38. *(Preliminaries for the International Math. Olympiad, Romania)* The set $M = \{1, 2, \ldots, 2n\}$ is partitioned into k sets $M_1, \ldots, M_k$, where $n \geq k^3 + k$. Show that there exist $i, j \in \{1, \ldots, k\}$ for which we can find $k + 1$ distinct even numbers $2r_1, \ldots, 2r_{k+1} \in M_i$ with the property that $2r_1 - 1, \ldots, 2r_{k+1} - 1 \in M_j$.

Problem 2.39. *(Preliminaries for the International Math. Olympiad, Great Britain)* Let S be the set of odd integers not divisible by 5 and smaller than $30m$, where $m \in \mathbb{Z}_+^*$. Find the smallest k such that every subset $A \subset S$ of k elements contains two distinct integers, one of which divides the other.

Problem 2.40. Prove that $\prod_{1 \leq j < i \leq n} \frac{a_i - a_j}{i - j}$ is a natural number whenever $a_1 \leq a_2 \leq \cdots \leq a_n$ are integers.

Problem 2.41. Is there an infinite set $A \subset \mathbb{Z}_+$ such that for all $x, y \in A$ neither x nor $x + y$ is a perfect power, i.e., a^k, for $k \geq 2$? More generally, is there an infinite set $A \subset \mathbb{Z}_+$ such that for any nonempty finite collection $x_i \in A, i \in J$, the sum $\sum_{i \in J} x_i$ is not a perfect power?

Problem 2.42. *(Amer. Math. Monthly)* Let

$$f(n) = \max A_1^{A_2^{A_3^{\cdot^{\cdot^{\cdot^{A_k}}}}}},$$

where $n = A_1 + \cdots + A_k$. Thus, $f(1) = 1$, $f(2) = 2$, $f(3) = 3$, $f(4) = 4$, $f(5) = 9$, $f(6) = 27$, $f(7) = 512$, etc. Determine $f(n)$.

Problem 2.43. Consider a set M with m elements and $A_1, \ldots, A_n$ distinct subsets of M such that $\operatorname{card}(A_i \cap A_j) = r \geq 1$ for all $1 \leq i \neq j \leq n$. Prove that $n \leq m$.

Problem 2.44. *(Amer. Math. Monthly)* Set $\pi(n)$ for the number of prime numbers less than or equal to n. Prove that there are at most $\pi(n)$ numbers $1 < a_1 < \cdots < a_k \leq n$ with $\gcd(a_i, a_j) = 1$.

Problem 2.45. *(Nieuw Archief v. Wiskunde)* Prove that for every k, there exists n such that the nth term of the Fibonacci sequence F_n is divisible by k. Recall that F_n is determined by the recurrence: $F_{n+2} = F_n + F_{n+1}$, for $n \geq 0$, where the first terms are $F_0 = 0$, $F_1 = 1$.

Problem 2.46. *(Amer. Math. Monthly)* Consider a set of consecutive integers $C + 1, C + 2, \ldots, C + n$, where $C > n^{n-1}$. Show that there exist distinct prime numbers $p_1, p_2, \ldots, p_n$ such that $C + j$ is divisible by p_j.

Problem 2.47. *(Nieuw Archief v. Wiskunde)* Let p be a prime number and $f(p)$ the smallest integer for which there exists a partition of the set $\{2, 3, \ldots, p\}$ into $f(p)$ classes such that whenever $a_1, \ldots, a_k$ belong to the same class of the partition, the equation

$$\sum_{i=1}^{k} x_i a_i = p$$

does not have solutions in nonnegative integers. Estimate $f(p)$.

Problem 2.48. *(Schweitzer Competition)* Consider m distinct natural numbers a_i smaller than N such that $\operatorname{lcm}(a_i, a_j) \leq N$ for all i, j. Prove that $m \leq 2\left[\sqrt{N}\right]$.

Problem 2.49. The set $M \subset \mathbb{Z}_+$ is called A-sum-free, where $A = (a_1, a_2, \ldots, a_k) \in \mathbb{Z}_+^k$, if for any choice of $x_1, x_2, \ldots, x_k \in M$ we have $a_1 x_1 + a_2 x_2 + \cdots + a_k x_k \notin M$. If A, B are two vectors, we define $f(n; A, B)$ as the greatest number h such that there exists a partition of the set of consecutive integers $\{n, n + 1, \ldots, h\}$ into S_1 and S_2 such that S_1 is A-sum-free and S_2 is B-sum-free. Assume that $B = (b_1, b_2, \ldots, b_m)$ and that the conditions below are satisfied:

$$a_1 + a_2 + \cdots + a_k = b_1 + b_2 + \cdots + b_m = s,$$

and

$$\min_{1 \leq j \leq k} a_j = \min_{1 \leq j \leq m} b_j = 1, \quad k, m \geq 2.$$

Prove that $f(n; A, B) = ns^2 + n(s - 1) - 1$.

Problem 2.50. *(Amer. Math. Monthly)* Let $1 \leq a_1 < a_2 < \cdots < a_n < 2n$ be a sequence of natural numbers for $n \geq 6$. Prove that

$$\min_{i,j} \operatorname{lcm}(a_i, a_j) \leq 6\left(\left[\frac{n}{2}\right] + 1\right).$$

Moreover, the constant 6 is sharp.

Problem 2.51. *(Amer. Math. Monthly)* Let $1 \leq a_1 < a_2 < \cdots < a_k < n$ be such that $\gcd(a_i, a_j) \neq 1$ for all $1 \leq i < j \leq k$. Determine the maximum value of k.

Problem 2.52. *(International Math. Olympiad 1978)* Consider the increasing sequence $f(n) \in \mathbb{Z}_+, 0 < f(1) < f(2) < \cdots < f(n) < \cdots$. It is known that the nth element in increasing order among the positive integers that are not terms of this sequence is $f(f(n)) + 1$. Find the value of $f(240)$.

Problem 2.53. *(Amer. Math. Monthly)* We define inductively three sequences of integers $(a_n), (b_n), (c_n)$ as follows:

1. $a_1 = 1$, $b_1 = 2$, $c_1 = 4$;
2. a_n is the smallest integer that does not belong to the set
   $\{a_1, \ldots, a_{n-1}, b_1, \ldots, b_{n-1}, c_1, \ldots, c_{n-1}\}$;
3. b_n is the smallest integer that does not belong to the set
   $\{a_1, \ldots, a_{n-1}, a_n, b_1, \ldots, b_{n-1}, c_1, \ldots, c_{n-1}\}$;
4. $c_n = 2b_n + n - a_n$.

Prove that
$$0 < n\left(1 + \sqrt{3}\right) - b_n < 2 \quad \text{for all } n \in \mathbb{Z}_+.$$

2.3 Geometric Combinatorics

Problem 2.54. We consider n points in the plane that determine C_n^2 segments, and to each segment one associates either $+1$ or -1. A triangle whose vertices are among these points will be called negative if the product of numbers associated to its sides is negative. Show that if n is even, then the number of negative triangles is even. Moreover, for odd n, the number of negative triangles has the same parity as the number p of segments labeled -1.

Problem 2.55. *(Amer. Math. Monthly)* Given n find a finite set S consisting of natural numbers larger than n with the property that for any $k \geq n$ the $k \times k$ square can be tiled by a family of $s_i \times s_i$ squares, where $s_i \in S$.

Problem 2.56. We consider $3n$ points $A_1, \ldots, A_{3n}$ in the plane whose positions are defined recursively by means of the following rule: first, the triangle $A_1 A_2 A_3$ is equilateral; further, the points A_{3k+1}, A_{3k+2}, and A_{3k+3} are the midpoints of the sides of the triangle $A_{3k} A_{3k-1} A_{3k-2}$. Let us assume that the $3n$ points are colored with two colors. Show that for $n \geq 7$ there exists at least one isosceles trapezoid having vertices of the same color.

Problem 2.57. Is there a coloring of all lattice points in the plane using only two colors such that there are no rectangles with all vertices of the same color whose side ratio belongs to $\{1, \frac{1}{2}, \frac{1}{3}, \frac{2}{3}\}$?

Problem 2.58. *(Amer. Math. Monthly)* Let G be a planar graph and let P be a path in G. We say that P has a (transversal) self-intersection in the vertex v if the path has a (transversal) self-intersection from the curve-theoretic viewpoint. Let us give an example: Take the point 0 in the plane and the segments $01, 02, 03, 04$ going counterclockwise around 0. Then a path traversing first 103 and then 204 has a (transversal) self-intersection at 0, while a path going first along 102 and further on 304 does not have a (transversal) self intersection.

Prove that any connected planar graph G with only even-degree vertices admits an Eulerian circuit without self-intersections. Recall that an Eulerian circuit is a path along the edges of the graph that passes precisely once along each edge of the graph.

Problem 2.59. *(Hungarian Competition)* Let us consider finitely many points in the plane that are not all collinear. Assume that one associates to each point a number from the set $\{-1, 0, 1\}$ such that the following property holds: for any line determined by two points from the set, the sum of numbers associated to all points lying on that line equals zero. Show that, if the number of points is at least three, then to each point one is associated with 0.

Problem 2.60. *(Amer. Math. Monthly)* If one has a set of squares with total area smaller than 1, then one can arrange them inside a square of side length $\sqrt{2}$, without any overlaps.

Problem 2.61. Prove that for each k there exist k points in the plane, no three collinear and having integral distance from each other. If we have an infinite set of points with integral distances from each other, then all points are collinear.

Problem 2.62. *(International Math. Olympiad 1984)* Let O, A be distinct points in the plane. For each point x in the plane, we write $\alpha(x) = \widehat{xOA}$ (counterclockwise). Let $C(x)$ be the circle of center O and radius $|Ox| + \frac{\alpha(x)}{|Ox|}$. If the points in the plane are colored with finitely many colors, then there exists a point y with $\alpha(y) > 0$ such that the color of y also belongs to the circle $C(y)$.

Problem 2.63. *(Amer. Math. Monthly)* Let $k, n \in \mathbb{Z}_+$.

1. Assume that $n - 1 \leq k \leq \frac{n(n-1)}{2}$. Show that there exist n distinct points $x_1, \ldots, x_n$ on a line that determine exactly k distinct distances $|x_i - x_j|$.
2. Suppose that $[\frac{n}{2}] \leq k \leq \frac{n(n-1)}{2}$. Then there exist n points in the plane that determine exactly k distinct distances.
3. Prove that for any $\varepsilon > 0$, there exists some constant $n_0 = n_0(\varepsilon)$ such that for any $n > n_0$ and $\varepsilon n < k < \frac{n(n-1)}{2}$, there exist n points in the plane that determine exactly k distinct distances.

Problem 2.64. *(International Math. Olympiad 1978)* Show that it is possible to pack $2n(2n + 1)$ nonoverlapping pieces having the form of a parallelepiped of dimensions $1 \times 2 \times (n + 1)$ in a cubic box of side $2n + 1$ if and only if n is even or $n = 1$.

Problem 2.65. *(Putnam Competition 1964)* Let F be a finite subset of $\mathbb{R}$ with the property that any value of the distance between two points from F (except for the largest one) is attained at least twice, i.e., for two distinct pairs of points. Prove that the ratio of any two distances between points of F is a rational number.

3

Geometry

3.1 Synthetic Geometry

Problem 3.1. Let I be the center of the circle inscribed in the triangle ABC and consider the points α, β, γ situated on the perpendiculars from I on the sides of the triangle ABC such that

$$|I\alpha| = |I\beta| = |I\gamma|.$$

Prove that the lines $A\alpha$, $B\beta$, $C\gamma$ are concurrent.

Problem 3.2. We consider the angle xOy and a point $A \in Ox$. Let (C) be an arbitrary circle that is tangent to Ox and Oy at the points H and D, respectively. Set AE for the tangent line drawn from A to the circle (C) that is different from AH. Show that the line DE passes through a fixed point that is independent of the circle (C) chosen above.

Problem 3.3. Let C be a circle of center O and A a fixed point in the plane. For any point $P \in C$, let M denote the intersection of the bisector of the angle $\widehat{AOP}$ with the circle circumscribed about the triangle AOP. Find the geometric locus of M as P runs over the circle C.

Problem 3.4. Let ABC be an isosceles triangle having $|AB| = |AC|$. If AS is an interior Cevian that intersects the circle circumscribed about ABC at S, then describe the geometric locus of the center of the circle circumscribed about the triangle BST, where $\{T\} = AS \cap BC$.

Problem 3.5. Let AB, CD, EF be three chords of length one on the unit circle. Then the midpoints of the segments $|BC|$, $|DE|$, and $|AF|$ form an equilateral triangle.

Problem 3.6. *(Amer. Math. Monthly)* Denote by P the set of points of the plane. Let $\star : P \times P \to P$ be the following binary operation: $A \star B = C$, where C is the unique point in the plane such that ABC is an oriented equilateral triangle whose orientation is counterclockwise. Show that $\star$ is a nonassociative and noncommutative operation satisfying the following "medial property":

$$(A \star B) \star (C \star D) = (A \star C) \star (B \star D).$$

Problem 3.7. Consider two distinct circles C_1 and C_2 with nonempty intersection and let A be a point of intersection. Let $P, R \in C_1$ and $Q, S \in C_2$ be such that PQ and RS are the two common tangents. Let U and V denote the midpoints of the chords PR and QS. Prove that the triangle AUV is isosceles.

Problem 3.8. *(Amer. Math. Monthly)* If the planar triangles AUV, VBU, and UVC are directly similar to a given triangle, then so is ABC. Recall that two triangles are directly similar if one can obtain one from the other using a homothety with positive ratio, rotations and translations.

Problem 3.9. *(Amer. Math. Monthly)* Find, using a straightedge and a compass, the directrix and the focus of a parabola. Recall that the parabola is the geometric locus of those points P in the plane that are at equal distance from a point O (called the focus) and a line d called the directrix.

Problem 3.10. Prove that if M is a point in the interior of a circle and $AB \perp CD$ are two chords perpendicular at M, then it is possible to construct an inscribable quadrilateral with the following lengths:

$$\big| |AM| - |MB| \big|, \ |AM| + |MB|, \ \big| |DM| - |MC| \big|, \ |DM| + |MC|.$$

Problem 3.11. *(Amer. Math. Monthly)* If the Euler line of a triangle passes through the Fermat point, then the triangle is isosceles.

Problem 3.12. Consider a point M in the interior of the triangle ABC, and choose $A' \in AM$, $B' \in BM$, and $C' \in CM$. Let P, Q, R, S, T, and U be the intersections of the sides of ABC and $A'B'C'$. Show that PS, TQ, and RU meet at M.

Problem 3.13. Show that if an altitude in a tetrahedron crosses two other altitudes, then all four altitudes are concurrent.

Problem 3.14. Three concurrent Cevians in the interior of the triangle ABC meet the corresponding opposite sides at A_1, B_1, C_1. Show that their common intersection point is uniquely determined if $|BA_1|, |CB_1|$, and $|AC_1|$ are equal.

Problem 3.15. *(International Math. Olympiad 1983)* Let $ABCD$ be a convex quadrilateral with the property that the circle of diameter AB is tangent to the line CD. Prove that the circle of diameter CD is tangent to the line AB if and only if AD is parallel to BC.

Problem 3.16. Let A', B', C' be points on the sides BC, CA, AB of the triangle ABC. Let M_1, M_2 be the intersections of the circle $A'B'C'$ with the circle ABA' and let N_1, N_2 be the analogous intersections of the circle $A'B'C'$ with the circle ABB'.

1. Prove that M_1M_2, N_1N_2, AB are either parallel or concurrent, in a point that we denote by A_1';

2. Prove that the analogously defined points A_1', B_1', C_1' are collinear.

Problem 3.17. *(Putnam Competition)* A circumscribable quadrilateral of area $S = \sqrt{abcd}$ is inscribable.

Problem 3.18. *(Nieuw Archief v. Wiskunde)* Let O be the center of the circumcircle, Ge the Gergonne point, Na the Nagel point, and G_1, N_1, the isogonal conjugates of G and N, respectively. Prove that G_1, N_1, and O are collinear (see also the Glossary for definitions of the important points in a triangle).

3.2 Combinatorial Geometry

Problem 3.19. Consider a rectangular sheet of paper. Prove that given any $\varepsilon > 0$, one can use finitely many foldings of the paper along its sides in either 2 equal parts or 3 equal parts to obtain a rectangle whose sides are in ratio r for some r satisfying $1 - \epsilon \leq r \leq 1 + \epsilon$.

Problem 3.20. *(Amer. Math. Monthly)* Show that there exist at most three points on the unit circle with the distance between any two being greater than $\sqrt{2}$.

Problem 3.21. *(Komal)* A convex polygon with $2n$ sides has at least n diagonals not parallel to any of its sides.

Problem 3.22. Let d be the sum of the lengths of the diagonals of a convex polygon $P_1 \ldots P_n$ and p its perimeter. Prove that for $n \geq 4$, we have

$$n - 3 < 2\frac{d}{p} < \left[\frac{n}{2}\right]\left[\frac{n+1}{2}\right] - 2.$$

Problem 3.23. *(Amer. Math. Monthly)* Find the convex polygons with the property that the function $D(p)$, which is the sum of the distances from an interior point p to the sides of the polygon, does not depend on p.

Problem 3.24. Prove that a sphere of diameter 1 cannot be covered by n strips of width l_i if $\sum_{i=1}^{n} l_i < 1$. Prove that a circle of diameter 1 cannot be covered by n strips of width l_i if $\sum_{i=1}^{n} l_i < 1$.

Problem 3.25. *(Kvant)* Consider n points lying on the unit sphere. Prove that the sum of the squares of the lengths of all segments determined by the n points is less than n^2.

Problem 3.26. *(Kvant)* The sum of the vectors $\overrightarrow{OA_1}, \ldots, \overrightarrow{OA_n}$ is zero, and the sum of their lengths is d. Prove that the perimeter of the polygon $A_1 \ldots A_n$ is greater than $4d/n$.

Problem 3.27. *(Amer. Math. Monthly)* Find the largest numbers a_k, for $1 \leq k \leq 7$, with the property that for any point P lying in the unit cube with vertices $A_1 \ldots A_8$, at least k among the distances $|PA_j|$ to the vertices are greater or equal than a_k.

Problem 3.28. *(Amer. Math. Monthly)* The line determined by two points is said to be admissible if its slope is equal to 0, 1, −1, or ∞. What is the maximum number of admissible lines determined by n points in the plane?

Problem 3.29. If $A = \{z_1, \ldots, z_n\} \subset \mathbb{C}$, then there exists a subset $B \subset A$ such that

$$\left| \sum_{z \in B} z \right| \geq \pi^{-1} \sum_{i=1}^{n} |z_i|.$$

Problem 3.30. *(Amer. Math. Monthly)* Let $A_1, \ldots, A_n$ be the vertices of a regular n-gon inscribed in a circle of center O. Let B be a point on the arc of circle $\overarc{A_1 A_n}$ and set $\theta = \widehat{A_n O B}$. If we set $a_k = |B A_k|$, then find the sum

$$\sum_{k=1}^{n} (-1)^k a_k$$

in terms of θ.

Problem 3.31. *(Amer. Math. Monthly)* Consider n distinct complex numbers $z_i \in \mathbb{C}$ such that

$$\min_{i \neq j} |z_i - z_j| \geq \max_{i \leq n} |z_i|.$$

What is the greatest possible value of n?

Problem 3.32. The interior of a triangle can be tiled by $n \geq 9$ pentagonal convex surfaces. What is the minimal value of n such that a triangle can be tiled by n hexagonal strictly convex surfaces?

Problem 3.33. *(Putnam Competition)* We say that a transformation of the plane is a congruence if it preserves the length of segments. Two subsets are congruent if there exists a congruence sending one subset onto the other. Show that the unit disk cannot be partitioned into two congruent subsets.

Problem 3.34. Prove that the unit disk cannot be partitioned into two subsets of diameter strictly smaller than 1, where the diameter of a set is the supremum distance between two of its points.

Problem 3.35. A continuous planar curve L has extremities A and B at distance $|AB| = 1$. Show that for any natural number n there exists a chord determined by two points $C, D \in L$ that is parallel to AB and whose length $|CD|$ equals $\frac{1}{n}$.

Problem 3.36. The diameter of a set is the supremum of the distance between two of its points. Prove that any planar set of unit diameter can pe partitioned into three parts of diameter no more than $\frac{\sqrt{3}}{2}$.

Problem 3.37.   1. Prove that a finite set of n points in $\mathbb{R}^3$ of unit diameter can be covered by a cube of side length $1 - \frac{2}{3n(n-1)}$.

2. Prove that any planar set of n points having unit diameter can pe partitioned into three parts of diameter less than $\frac{\sqrt{3}}{2} \cos \frac{2\pi}{3n(n-1)}$.

Problem 3.38. Prove that any convex set in $\mathbb{R}^n$ of unit diameter having a smooth boundary can be partitioned into $n + 1$ parts of diameter $d < 1$.

Problem 3.39. Let D be a convex body in $\mathbb{R}^3$ and let $\sigma(D) = \sup_\pi \text{area}(\pi \cap D)$, where the supremum is taken over all positions of the variable plane π. Prove that D can be divided into two parts D_1 and D_2 such that $\sigma(D_i) < \sigma(D)$.

Problem 3.40. If we have k vectors $v_1, v_2, \ldots, v_k$ in $\mathbb{R}^n$ and $k \le n + 1$, then there exist two vectors making an angle θ with $\cos \theta \ge -\frac{1}{k-1}$. Equality holds only when the endpoints of the vectors form a regular $(k - 1)$-simplex.

Problem 3.41. 1. Consider a finite family of bounded closed convex sets in the plane such that any three members of the family have nonempty intersection. Prove that the intersection of all members of the family is nonempty.

2. A set of unit diameter in $\mathbb{R}^2$ can be covered by a ball of radius $\sqrt{\frac{1}{3}}$.

Problem 3.42. *(Amer. Math. Monthly)* Let T be a right isosceles triangle. Find the disk D such that the difference between the areas of $T \cup D$ and $T \cap D$ is minimal.

Problem 3.43. *(Amer. Math. Monthly)* Let r be the radius of the incircle of an arbitrary triangle lying in the closed unit square. Prove that $r \le \frac{\sqrt{5}-1}{4}$.

Problem 3.44. Let P be a point in the interior of the tetrahedron $ABCD$, with the property that $|PA| + |PB| + |PC| + |PD|$ is minimal. Prove that $\widehat{APB} = \widehat{CPD}$ and that these angles have a common bisector.

Problem 3.45. *(Putnam Competition 1948)* Let $OA_1, \ldots, OA_n$ be n linearly independent vectors of lengths $a_1, \ldots, a_n$. We construct the parallelepiped H having these vectors as sides. Then consider the n altitudes in H as a new set of vectors and further, construct the parallelepiped E associated with the altitudes. If h is the volume of H and e the volume of E, then prove that

$$he = (a_1 \ldots a_n)^2.$$

Problem 3.46. Let F be a symmetric convex body in $\mathbb{R}^3$ and let $A_{F,\lambda}$ denote the family of all sets homothetic to F in the ratio λ that have only boundary points in common with F. Set $h_F(\lambda)$ for the greatest integer k such that $A_{F,\lambda}$ contains k sets with pairwise disjoint interiors. Prove that

$$h_F(\lambda) \le \frac{(1 + 2\lambda)^3 - 1}{\lambda^3}.$$

Problem 3.47. Let Δ denote the square of equations $|x_i| \leq 1$, $i = 1, 2$, in the plane, and let $A = (a_1|a_2)$ be an arbitrary nonsingular 2×2 matrix partitioned into two columns. We identify each column with a vector in $\mathbb{R}^2$. Prove that the following inequality holds:

$$\min_A \max_{x \in \Delta} \left| \frac{\langle a_1, x \rangle \langle a_2, x \rangle}{\det A} \right| = \frac{1}{2},$$

where $\langle x, y \rangle = x_1 x_2 + y_1 y_2$ is the usual scalar product.

Problem 3.48. We denote by $\delta(r)$ the minimal distance between a lattice point and the circle $C(O, r)$ of radius r centered at the origin O of the coordinate system in the plane. Prove that

$$\lim_{r \to \infty} \delta(r) = 0.$$

Problem 3.49. *(Putnam Competition)* Consider a curve C of length l that divides the surface of the unit sphere into two parts of equal area. Show that $l \geq 2\pi$.

Problem 3.50. *(Amer. Math. Monthly)* Let K be a planar closed curve of length 2π. Prove that K can be inscribed in a rectangle of area 4.

Problem 3.51.   1. Consider a family of plane convex sets with area a, perimeter p, and diameter d. If the family covers area A, then there exists a subfamily with pairwise disjoint interiors that covers at least area λA, where $\lambda = \frac{a}{a + pd + \pi d^2}$.
  2. Assume that any two members of the family have nonempty intersection. Prove that there exists then a subfamily with pairwise disjoint interiors that covers area at least μA, where $\mu = \frac{a}{\pi d^2}$.

Problem 3.52. *(Putnam Compettion 1957)* Let C be a regular polygon with k sides. Prove that for every n there exists a planar set $S(n) \subset \mathbb{R}^2$ such that any subset consisting of n points of $S(n)$ can be covered by C, but $S(n)$ itself cannot be included in C.

Problem 3.53. Let M be a convex polygon and let $S_1, \ldots, S_n$ be pairwise disjoint disks situated in the interior of M. Does there exist a partition $M = D_1 \cup \cdots \cup D_n$ such that D_i are convex disjoint polygons, each of which contains precisely one disk?

Problem 3.54. Consider an inscribable n-gon partitioned by means of $n - 2$ nonintersecting diagonals into $n - 2$ triangles. Prove that the sum of the radii of the circles inscribed in these triangles does not depend on the particular partition.

Problem 3.55. *(Elemente der Mathematik)* Prove that in an ellipse having axes of lengths a and b and total length L, we have $L > \pi(a + b)$.

Problem 3.56. *(Kvant)* Let F be a convex planar domain and F' denote its image by a homothety of ratio $-\frac{1}{2}$. Is it true that one can translate F' in order for it to be contained in F? Can the constant $\frac{1}{2}$ be improved? Generalize to n dimensions.

Problem 3.57. *(Nieuw Archief v. Wiskunde)* A classical theorem, due to Cauchy, states that a strictly convex polyhedron in $\mathbb{R}^3$ whose faces are rigid must be globally rigid. Here, rigidity means continuous rigidity in the sense that any continuous deformation of the polyhedron in $\mathbb{R}^3$ that keeps the lengths of edges fixed is the restriction of a deformation of rigid Euclidean motions of three-space. Prove that a 3-dimensional cube immersed in $\mathbb{R}^n$ remains rigid for all $n > 3$.

Problem 3.58. Consider finitely many great circles on a sphere such that not all of them pass through the same point. Show that there exists a point situated on exactly two circles. Deduce that if we have a set of n points in the plane, not all of them lying on the same line, then there must exist one line passing through precisely two points of the given set.

Problem 3.59. Given a finite set of points in the plane labeled with $+1$ or -1, and not all of them collinear, show that there exists a line determined by two points in the set such that all points of the set lying on that line are of the same sign.

Problem 3.60. *(Amer. Math. Monthly)* If Q is a given rectangle and $\varepsilon > 0$, then Q can be covered by the union of a finite collection S of rectangles with sides parallel to those of Q in such a way that the union of every nonoverlapping subcollection of S has area less than ε.

Problem 3.61. Prove that the 3-dimensional ball cannot be partitioned into three sets of strictly smaller diameter.

3.3 Geometric Inequalities

Problem 3.62. If a, b, c, r, R are the usual notations in the triangle, show that

$$\frac{1}{2rR} \leq \frac{1}{3}\left(\sum \frac{1}{a}\right)^2 \leq \sum \frac{1}{a^2} \leq \frac{1}{4r^2}.$$

Problem 3.63. *(Amer. Math. Monthly)* If a, b, c are the sides of a triangle, then prove that $\frac{(b+c)^2}{4bc} \leq \frac{m_a}{w_a}$ and $\frac{b^2+c^2}{2bc} \leq \frac{m_a}{k_a}$, where m_a, w_a, k_a denote respectively the lengths of the median, bisector, and altitude issued from A.

Problem 3.64. *(Putnam Competition)* If $S(x, y, z)$ is the area of a triangle with sides x, y, z, prove that

$$\sqrt{S(a, b, c)} + \sqrt{S(a', b', c')} \leq \sqrt{S(a + a', b + b', c + c')}.$$

Problem 3.65. *(Nieuw Archief v. Wiskunde)* It is known that in any triangle we have the inequality

$$3\sqrt{3}r \leq p \leq 2R + (3\sqrt{3} - 4)r,$$

where p denotes the semiperimeter. Prove that in an obtuse triangle we have

$$(3 + 2\sqrt{2})r < p < 2R + r.$$

Problem 3.66. Prove the Euler inequality

$$R \geq 2r.$$

Problem 3.67. Prove that in a triangle we have the inequalities

$$36r^2 \leq a^2 + b^2 + c^2 \leq 9R^2.$$

Problem 3.68. *(Amer. Math. Monthly)*

1. Let ABC and $A'B'C'$ be two triangles. Prove that

$$\frac{a^2}{a'} + \frac{b^2}{b'} + \frac{c^2}{c'} \leq R^2 \frac{(a' + b' + c')^2}{a'b'c'}.$$

2. Derive that

$$a^2 + b^2 + c^2 \leq 9R^2,$$

$$\cos A \cos B \cos C \leq \frac{1}{8}.$$

Problem 3.69. *(Elemente der Mathematik)* Prove that the following inequalities hold in a triangle:

$$4 \sum_{\text{cyclic}} h_A h_B \leq 12S\sqrt{3} \leq 54Rr \leq 3 \sum_{\text{cyclic}} ab \leq 4 \sum_{\text{cyclic}} r_A r_B.$$

Problem 3.70. *(Amer. Math. Monthly)* Prove that in an any triangle ABC, we have

$$\frac{\sqrt{1 + 8\cos^2 B}}{\sin A} + \frac{\sqrt{1 + 8\cos^2 C}}{\sin B} + \frac{\sqrt{1 + 8\cos^2 A}}{\sin C} \geq 6.$$

Problem 3.71. Let P be a point in the interior of the triangle ABC. We denote by R_a, R_b, R_c the distances from P to A, B, C and by r_a, r_b, r_c the distances to the sides BC, CA, AB. Prove that

$$\sum_{\text{cyclic}} R_a^2 \sin^2 A \leq 3 \sum_{\text{cyclic}} r_a^2,$$

with equality if and only if P is the Lemoine point (i.e., the symmedian point).

Problem 3.72. Prove the inequalities

$$16Rr - 5r^2 \leq p^2 \leq 4R^2 + 4Rr + 3r^2.$$

Problem 3.73. Prove the following inequalities, due to Roché:

$$2R^2 + 10Rr - r^2 - 2(R - 2r)\sqrt{R^2 - 2Rr}$$
$$\leq p^2 \leq 2R^2 + 10Rr - r^2 + 2(R - 2r)\sqrt{R^2 - 2Rr}.$$

4

Analysis

Problem 4.1. *(Amer. Math. Monthly)* Prove that $z \in \mathbb{C}$ satisfies $|z| - \Re z \leq \frac{1}{2}$ if and only if $z = ac$, where $|\bar{c} - a| \leq 1$. We denote by $\Re z$ the real part of the complex number z.

Problem 4.2. Let $a, b, c \in \mathbb{R}$ be such that $a + 2b + 3c \geq 14$. Prove that $a^2 + b^2 + c^2 \geq 14$.

Problem 4.3. Let $f_n(x)$ denote the Fibonacci polynomial, which is defined by

$$f_1 = 1, \quad f_2 = x, \quad f_n = x f_{n-1} + f_{n-2}.$$

Prove that the inequality

$$f_n^2 \leq (x^2 + 1)^2 (x^2 + 2)^{n-3}$$

holds for every real x and $n \geq 3$.

Problem 4.4. *(Amer. Math. Monthly)* Prove the inequality

$$\min\left((b - c)^2, (c - a)^2, (a - b)^2\right) \leq \frac{1}{2}\left(a^2 + b^2 + c^2\right).$$

Generalize to $\min_{1 \leq k < i \leq n}(a_k - a_i)^2$.

Problem 4.5. Let $a_1, \ldots, a_n$ be the lengths of the sides of a polygon and P its perimeter. Then

$$\frac{a_1}{P - a_1} + \frac{a_2}{P - a_2} + \cdots + \frac{a_n}{P - a_n} \geq \frac{n}{n - 1}.$$

Problem 4.6. Assume that $a, b,$ and c are positive numbers that cannot be the sides of a triangle. Then the following inequality holds:

$$(abc)^2(a + b + c)^2(a + b - c)(a - b + c)(b - a + c)$$
$$\geq \left(a^2 + b^2 + c^2\right)^3 \left(a^2 + b^2 - c^2\right)\left(a^2 - b^2 + c^2\right)(b^2 - a^2 + c^2).$$

Problem 4.7. *(Amer. Math. Monthly)* Prove that $\sin^2 x < \sin x^2$, for $0 < x \leq \sqrt{\frac{\pi}{2}}$.

Problem 4.8. Let $P(x) = a_0 + a_1 x + \cdots + a_n x^n$ be a real polynomial of degree $n \geq 2$ such that

$$0 < a_0 < -\sum_{j=1}^{[\frac{n}{2}]} \frac{1}{2k+1} a_{2k}.$$

Show that $P(x)$ has a real zero in $(-1, 1)$.

Problem 4.9. *(Amer. Math. Monthly)* Find the minimum of β and the maximum of α for which

$$\left(1 + \frac{1}{n}\right)^{n+\alpha} \leq e \leq \left(1 + \frac{1}{n}\right)^{n+\beta}$$

holds for all $n \in \mathbb{Z}_+$.

Problem 4.10. *(International Math. Olympiad 1984)* Prove that for nonnegative x, y, and z such that $x + y + z = 1$, the following inequality holds:

$$0 \leq xy + yz + zx - 2xyz \leq \frac{7}{27}.$$

Problem 4.11. *(Putnam Competition 1948)* Consider the sequence of nonzero complex numbers $a_1, \ldots, a_n, \ldots$ with the property that $|a_r - a_s| > 1$ for $r \neq s$. Prove that

$$\sum_{n=1}^{\infty} \frac{1}{a_n^3}$$

converges.

Problem 4.12. *(Amer. Math. Monthly)* Consider the sequence S_n given by

$$S_n = \frac{n+1}{2^{n+1}} \sum_{i=1}^{n} \frac{2^i}{i}.$$

Find $\lim_{n \to \infty} S_n$.

Problem 4.13. *(Putnam Competition 1951)* Prove that whenever $a, b > 0$ we have

$$\int_0^1 \frac{t^{a-1}}{1+t^b} \, dt = \frac{1}{a} - \frac{1}{a+b} + \frac{1}{a+2b} - \frac{1}{a+3b} + \cdots.$$

Problem 4.14. *(Putnam Competition 1951)* Let $a, b, c, d \in \mathbb{Z}_+^*$, and $r = 1 - \frac{a}{b} - \frac{c}{d}$. If $r > 0$ and $a + c \leq 1982$, then $r > \frac{1}{1983^3}$.

Problem 4.15. *(Amer. Math. Monthly)* Let $a_i \in \mathbb{R}$. Prove that

$$n \min(a_i) \leq \sum_{i=1}^{n} a_i - S \leq \sum_{i=1}^{n} a_i + S \leq n \max(a_i),$$

where $(n-1)S^2 = \sum_{1 \leq i < j \leq n} (a_i - a_j)^2$, with $S \neq 0$, and with equality if and only if $a_1 = \cdots = a_n$.

Problem 4.16. *(Schweitzer Competition)* Consider a periodic function $f : \mathbb{R} \to \mathbb{R}$, of period 1 that is nonnegative, concave on $(0, 1)$ and continuous at 0. Prove that for all real numbers x and natural numbers n the following inequality $f(nx) \leq nf(x)$ is satisfied.

Problem 4.17. *(Amer. Math. Monthly)* Let (a_n) be a sequence of positive numbers such that

$$\lim_{n \to \infty} \frac{(a_1 + \cdots + a_n)}{n} < \infty, \quad \lim_{n \to \infty} \frac{a_n}{n} = 0.$$

Does this imply that $\lim_{n \to \infty} \frac{a_1^2 + \cdots + a_n^2}{n^2} = 0$?

Problem 4.18. *(Amer. Math. Monthly)* Show that for a fixed $m \geq 2$, the following series converges for a single value of x:

$$S(x) = 1 + \ldots + \frac{1}{m-1} - \frac{x}{m} + \frac{1}{m+1} + \cdots + \frac{1}{2m-1} - \frac{x}{2m}$$
$$+ \frac{1}{2m+1} + \cdots + \frac{1}{3m-1} - \frac{x}{3m} + \cdots .$$

Problem 4.19. *(Amer. Math. Monthly)* Prove that if $z_i \in \mathbb{C}, 0 < |z_i| \leq 1$, then $|z_1 - z_2| \leq |\log z_1 - \log z_2|$.

Problem 4.20. *(Nieuw Archief v. Wiskunde)* The roots of the function of a complex variable

$$\zeta_4(s) = 1 + 2^s + 3^s + 4^s$$

are simple.

Problem 4.21. *(Putnam Competition 1953)* Compute

$$I = \int_0^{\pi/2} \log \sin x \, dx.$$

Problem 4.22. *(Putnam Competition 1957)* Let f a be positive, monotonically decreasing function on $[0, 1]$. Prove the following inequality:

$$\frac{\int_0^1 x f^2(x) dx}{\int_0^1 x f(x) dx} \leq \frac{\int_0^1 f^2(x) dx}{\int_0^1 f(x) dx}.$$

Problem 4.23. *(Schweitzer Competition)* Prove that every real number x such that $0 < x \leq 1$ can be represented as an infinite sum

$$x = \sum_{k=1}^{\infty} \frac{1}{n_k},$$

where the n_k are natural numbers such that $\frac{n_{k+1}}{n_k} \in \{2, 3, 4\}$.

Problem 4.24. *(Amer. Math. Monthly)* Let S_n denote the set of polynomials $p(z) = z^n + a_{n-1}z^{n-1} + \cdots + a_1 z + 1$, with $a_i \in \mathbb{C}$. Find

$$M_n = \min_{p \in S_n} \left(\max_{|z|=1} |p(z)| \right).$$

Problem 4.25. *(Amer. Math. Monthly)* Find an explicit formula for the value of F_n that is given by the recurrence $F_n = (n+2)F_{n-1} - (n-1)F_{n-2}$ and initial conditions $F_1 = a$, $F_2 = b$.

Problem 4.26. Find the positive functions $f(x, y)$ and $g(x, y)$ satisfying the following inequalities:

$$\left(\sum_{i=1}^{n} a_i b_i \right)^2 \le \left(\sum_{i=1}^{n} f(a_i, b_i) \right) \left(\sum_{i=1}^{n} g(a_i, b_i) \right) \le \left(\sum_{i=1}^{n} a_i^2 \right) \left(\sum_{i=1}^{n} b_i^2 \right)$$

for all $a_i, b_i \in \mathbb{R}$ and $n \in \mathbb{Z}_+$.

Problem 4.27. *(Putnam Competition 1946)* Let $g \in \mathcal{C}^1(\mathbb{R})$ be a smooth function such that $g(0) = 0$ and $|g'(x)| \le |g(x)|$. Prove that $g(x) = 0$.

Problem 4.28. *(Putnam Competition 1947)* Let $a_1, b_1, c_1 \in \mathbb{R}_+$ such that $a_1 + b_1 + c_1 = 1$ and define

$$a_{n+1} = a_n^2 + 2b_n c_n, \quad b_{n+1} = b_n^2 + 2a_n c_n, \quad c_{n+1} = c_n^2 + 2a_n b_n.$$

Prove that the sequences (a_n), (b_n), (c_n) have the same limit and find that limit.

Problem 4.29. *(Kvant)* Consider the sequence (a_n), given by $a_1 = 0$, $a_{2n+1} = a_{2n} = n - a_n$. Prove that $a_n = \frac{n}{3}$, for infinitely many values of n. Does there exist an n such that $|a_n - \frac{n}{3}| > 2005$? Also, prove that

$$\lim_{n \to \infty} \frac{a_n}{n} = \frac{1}{3}.$$

Problem 4.30. Compute the integral

$$f(a) = \int_0^1 \frac{\log(x^2 - 2x \cos a + 1)}{x} \, dx.$$

Problem 4.31. *(Amer. Math. Monthly)* Let $-1 < a_0 < 1$, and define $a_n = \left(\frac{1}{2}(1 + a_{n-1}) \right)^{1/2}$ for $n \ge 1$. Find the limits A, B, and C of the sequences

$$A_n = 4^n (1 - a_n), \quad B_n = a_1 \cdots a_n, \quad C_n = 4^n (B - a_1 a_2 \cdots a_n).$$

Problem 4.32. *(Amer. Math. Monthly)* Let consider the sequence given by the recurrence

$$a_1 = a, \quad a_n = a_{n-1}^2 - 2.$$

Determine those $a \in \mathbb{R}$ for which (a_n) is convergent.

Problem 4.33. Let $0 < a < 1$ and $I = (0, a)$. Find all functions $f : I \to \mathbb{R}$ satisfying at least one of the conditions below:

1. f is continuous and $f(xy) = xf(y) + yf(x)$.
2. $f(xy) = xf(x) + yf(y)$.

Problem 4.34. If $a, b, c, d \in \mathbb{C}, ac \neq 0$, prove that

$$\frac{\max(|ac|, |ad + bc|, |bd|)}{\max(|a|, |b|) \max(|c|, |d|)} \geq \frac{-1 + \sqrt{5}}{2}.$$

Problem 4.35. *(Putnam Competition)* Let $\sum_{i=1}^{\infty} x_i$ be a convergent series with decreasing terms $x_1 \geq x_2 \geq \cdots \geq x_n \geq \cdots > 0$ and let P be the set of numbers which can be written in the form $\sum_{i \in J} x_i$ for some subset $J \in \mathbb{Z}_+$. Prove that P is an interval if and only if

$$x_n \leq \sum_{i=n+1}^{\infty} x_i \text{ for every } n \in \mathbb{Z}_+.$$

Problem 4.36. *(Amer. Math. Monthly)* Does there exist a continuous function $f : (0, \infty) \to \mathbb{R}$ such that $f(x) = 0$ if and only if $f(2x) \neq 0$? What if we require only that f be continuous at infinitely many points?

Problem 4.37. *(Kvant)* Find the smallest number a such that for every real polynomial $f(x)$ of degree two with the property that $|f(x)| \leq 1$ for all $x \in [0, 1]$, we have $|f'(1)| \leq a$. Find the analogous number b such that $|f'(0)| \leq b$.

Problem 4.38. Let $f : \mathbb{R} \to \mathbb{R}$ be a function for which there exists some constant $M > 0$ satisfying

$$|f(x + y) - f(x) - f(y)| \leq M, \text{ for all } x, y \in \mathbb{R}.$$

Prove that there exists a unique additive function $g : \mathbb{R} \to \mathbb{R}$ such that

$$|f(x) - g(x)| \leq M, \text{ for all } x \in \mathbb{R}.$$

Moreover, if f is continuous, then g is linear.

Problem 4.39. *(Amer. Math. Monthly)* Show that if f is differentiable and if

$$\lim_{t \to \infty} \left(f(t) + f'(t) \right) = 1,$$

then

$$\lim_{t \to \infty} f(t) = 1.$$

Problem 4.40. *(Putnam Competition)* Let c be a real number, and let $f : \mathbb{R} \to \mathbb{R}$ be a smooth function of class C^3 such that $\lim_{x \to \infty} f(x) = c$ and $\lim_{x \to \infty} f'''(x) = 0$. Show that $\lim_{n \to \infty} f'(x) = \lim_{x \to \infty} f''(x) = 0$.

Problem 4.41. *(Putnam Competition)* Prove that the following integral equation has at most a continuous solution on $[0, 1] \times [0, 1]$:

$$f(x, y) = 1 + \int_0^x \int_0^y f(u, v) \, du \, dv.$$

Problem 4.42. *(Putnam Competition)* Find those $\lambda \in \mathbb{R}$ for which the functional equation

$$\int_0^1 \min(x, y) f(y) \, dy = \lambda f(x)$$

has a solution f that is nonzero and continuous on the interval $[0, 1]$. Find these solutions.

Problem 4.43. *(Amer. Math. Monthly)* Let X be an unbounded subset of the real numbers $\mathbb{R}$. Prove that the set

$$A_X = \{t \in \mathbb{R}; tX \text{ is dense modulo } 1\}$$

is dense in $\mathbb{R}$.

Problem 4.44. *(Putnam Competition)* Consider $P(z) = z^n + a_1 z^{n-1} + \cdots + a_n$, where $a_i \in \mathbb{C}$. If $|P(z)| = 1$ for all z satisfying $|z| = 1$, then $a_1 = \cdots = a_n = 0$.

Problem 4.45. Let $I \subset \mathbb{R}$ be an interval and $u, v : I \to \mathbb{R}$ smooth functions satisfying the equations

$$u''(x) + A(x)u(x) = 0, \quad v''(x) + B(x)v(x) = 0,$$

where A, B are continuous on I and $A(x) \geq B(x)$ for all $x \in I$. Assume that v is not identically zero. If $\alpha < \beta$ are roots of v, then there exists a root of u that lies within the interval (α, β), unless $A(x) = B(x)$, in which case u and v are proportional for $\alpha \leq x \leq \beta$.

Problem 4.46. *(Amer. Math. Monthly)* Let V be a finite-dimensional real vector space and $f : V \to \mathbb{R}$ a continuous mapping. For any basis $B = \{b_1, b_2, \ldots, b_n\}$ of V, consider the set

$$E_B = \{z_1 b_1 + \cdots + z_n b_n, \text{ where } z_i \in \mathbb{Z}\}.$$

Show that if f is bounded on E_B for any choice of the basis B, then f is bounded on V.

Problem 4.47. *(Amer. Math. Monthly)* It is known that if $f, g : \mathbb{C} \to \mathbb{C}$ are entire functions without common zeros then there exist entire functions $a, b : \mathbb{C} \to \mathbb{C}$ such that $a(z)f(z) + b(z)g(z) = 1$ for all $z \in \mathbb{C}$.

1. Prove that we can choose $a(z)$ without any zeros.
2. Is it possible to choose both a and b without zeros.

Problem 4.48. *(Amer. Math. Monthly)* Consider a compact set $X \subset \mathbb{R}$. Show that a necessary and sufficient condition for the existence of a monic nonconstant polynomial with real coefficients $h \in \mathbb{R}[x]$ such that $|h(x)| < 1$ for all $x \in X$ is the existence of monic nonconstant polynomial $g(x) \in \mathbb{R}[x]$ such that $|g(x)| < 2$ for all $x \in X$. Prove that 2 is the maximal number with this property.

Part II

Solutions and Comments to the Problems

Number Theory Solutions

Problem 1.1. Show that we have $C_n^k \equiv 0 \pmod 2$ for all k satisfying $1 \le k \le n-1$ if and only if $n = 2^\beta$, where $\beta \in \mathbb{Z}_+^*$. Here C_n^k denotes the number of combinations, i.e., the number of ways of picking up a subset of k elements from a set of n elements. Known also as the binomial coefficient or choice number and sometimes denoted as $\binom{n}{k}$ it is given by the formula

$$C_n^k = \binom{n}{k} = \frac{n!}{k!(n-k)!}$$

where the factorial $n!$ represents

$$n! = 1 \times 2 \times 3 \times \cdots \times n.$$

Solution 1.1. Denote by $\exp_k(m)$ the maximal exponent of m in k, i.e., the maximal r such that m^r divides k. Since $C_n^k = \frac{n!}{k!(n-k)!}$, the exponent of 2 in C_n^k has the value

$$\exp_{C_n^k}(2) = \exp_{n!}(2) - \exp_{k!}(2) - \exp_{(n-k)!}(2),$$

which is, by hypothesis, strictly positive for all considered k. Now, the exponent of 2 in a factorial is given by the following formula:

$$\exp_{m!}(2) = \sum_{i=1}^{\infty} \left[\frac{m}{2^i}\right],$$

where the brackets denote the integer part. This implies that our claim is equivalent to the inequality

$$\sum_{i=1}^{\infty} \left[\frac{n}{2^i}\right] - \sum_{i=1}^{\infty} \left[\frac{k}{2^i}\right] - \sum_{i=1}^{\infty} \left[\frac{n-k}{2^i}\right] > 0.$$

Let us now suppose that $n = 2^\beta + s$, with $1 \le s < 2^\beta$. If we take $k = s$, then the exponent above can be calculated as

$$\left(\left[\frac{2^\beta + s}{2}\right] + \cdots\right) - \left(\left[\frac{s}{2}\right] + \cdots\right) - \left(\left[\frac{2^\beta}{2}\right] + \cdots\right) = 0,$$

contradicting our assumptions. Thus, it is necessary that $n = 2^\beta$.

The sufficiency is established as follows. We have the identity $(a+b)^2 = a^2 + b^2$ in $\mathbb{Z}/2\mathbb{Z}$. Using induction on β, one proves that $(x + y)^{2^\beta} = x^{2^\beta} + y^{2^\beta} \in \mathbb{Z}/2\mathbb{Z}$. In particular, $(x + 1)^{2^\alpha} = x^{2^\alpha} + 1$, and therefore, by identifying the coefficients of the binomial expansion on the left- and right-hand sides, we obtain the claim.

Comments 1 *Using the formula $C_n^k = C_{n-1}^k + C_{n-1}^{k-1}$, we find that $C_n^k \equiv 1 \pmod 2$ for all $k \le n$ if and only if $n = 2^\alpha - 1$.*

Problem 1.2. Let $P = a_n x^n + \cdots + a_0$ be a polynomial with integer coefficients. Suppose that there exists a number p such that:

1. p does not divide a_n;
2. p divides a_i, for all $i \le n - 1$;
3. p^2 does not divide a_0.

Then P is an irreducible polynomial in $\mathbb{Z}[x]$.

Solution 1.2. Let $\varphi : \mathbb{Z} \to \mathbb{Z}/p\mathbb{Z}$ be the reduction modulo p and let $\varphi : \mathbb{Z}[x] \to \mathbb{Z}/p\mathbb{Z}[x]$ be the map consisting of taking the reduction modulo p of all coefficients. This map is actually a homomorphism of rings, meaning that it respects addition and multiplication of polynomials. We have

$$\varphi(P(x)) = \varphi(a_n)x^n \ne 0.$$

Assume that P is not irreducible, and so $P = P_1 \cdot P_2$. Then $\varphi(P) = \varphi(P_1) \cdot \varphi(P_2)$. Moreover, the only way to write $\varphi(a_n)x^n \in \mathbb{Z}/p\mathbb{Z}[x]$ as a product of two polynomials is $\beta x^k \cdot \gamma x^l$, where $k + l = n$ and $\beta\gamma = \varphi(a_n)$. In particular, if we write $P_1(x) = \beta_0 + \cdots + \beta_k x^k$, $P_2(x) = \gamma_0 + \cdots + \gamma_l x^l$, then $\varphi(\beta_k) = \beta$, $\varphi(\beta_i) = 0$ if $i < k$, and $\varphi(\gamma_i) = \gamma$, $\varphi(\gamma_i) = 0$ if $i < l$. Eventually, this yields $a_0 = \beta_0\gamma_0 \equiv 0 \pmod{p^2}$, which contradicts our assumptions.

Comments 2 *This result is known as the Eisenstein criterion. Using this result we can give an alternative solution to Problem 1.1 above. Assume that $C_n^k \equiv 0 \pmod 2$ for $k \in \{1, \ldots, n - 1\}$. Then the Eisenstein criterion shows that the polynomial $P(x) = (x + 1)^n + 1$ is irreducible. This means that $Q(x) = P(x - 1) = x^n + 1$ is equally irreducible. On the other hand, if we can write $n = 2^a(2b + 1)$, with $a, b \in \mathbb{Z}_+$, then $x^n + 1$ is divisible by $x^{2^a} + 1$; thus irreducibility of $Q(x)$ implies that $b = 0$. Therefore, $n = 2^a$, as claimed.*

Problem 1.3. Given $m_i, b_i \in \mathbb{Z}_+$, $i \in \{1, 2, \ldots, n\}$ such that $\gcd(m_i, m_j) = 1$ for all $i \ne j$, there exist integers x satisfying $x \equiv b_i \pmod{m_i}$ for all i. This result is usually known as the Chinese remainder theorem.

Solution 1.3. For $n = 2$, we have $m_1\mathbb{Z} + m_2\mathbb{Z} = \mathbb{Z}$. Therefore there exist $a_i \in m_i\mathbb{Z}$ such that $a_1 + a_2 = 1$. Then $x = a_1b_2 + a_2b_1$ satisfies the conditions of the statement.

Since m_i are relatively prime, for each $j \in \{1, 2, \ldots, n\}$ there exists $y_j \in \mathbb{Z}$ satisfying the conditions $y_j \equiv 1 \pmod{m_j}$ and $y_j \equiv 0 \pmod{m_1 \cdots m_{j-1}m_{j+1} \cdots m_n}$. Then take $x = b_1y_1 + \cdots + b_ny_n$, which satisfies the system of congruences.

Problem 1.4. If $\gcd(a, m) = 1$, then it is a classical result of Euler that we have the following congruence:
$$a^{\varphi(m)+1} \equiv 0 \pmod{m},$$
where $\varphi(m)$ is the Euler totient function, which counts how many positive integers smaller than m are relatively prime to m. Prove that this equality holds precisely for those numbers a, m such that for any prime number p that divides a, if p^k divides m then p^k also divides a.

Solution 1.4. Let p be a prime number and e the maximal exponent such that p^e divides m. If p divides a, then p^e divides a. If p does not divide a, then $a^{\varphi(p^e)} \equiv 1 \pmod{p^e}$, because $\varphi(m)$ is a multiple of $\varphi(p^e)$. Therefore $a^{\varphi(m)} \equiv 1 \pmod{p^e}$. This implies that p^e divides $a(a^{\varphi(m)} - 1)$. Hence m divides $a(a^{\varphi(m)} - 1)$.

In order to establish the converse, let p be a divisor of a such that p^k divides m but p^k does not divide a. Then p^k does not divide $a(a^{\varphi(m)} - 1)$, because obviously, p and $a^{\varphi(m)} - 1$ are relatively prime. Consequently, m is not a divisor of $a(a^{\varphi(m)} - 1)$.

Problem 1.5. Let $p > 2$ be a prime number and let $a_k \in \{0, 1, \ldots, p^2 - 1\}$ denote the value of k^p modulo p^2. Prove that
$$\sum_{k=1}^{p-1} a_k = \frac{p^3 - p^2}{2}.$$

Solution 1.5. We have
$$(p - k)^p = (-k)^p + C_p^1 p(-k)^{p-1} + \cdots + p^p \equiv -k^p \pmod{p^2}.$$

Therefore, $a_{p-k} = p^2 - a_k$. In particular, summing up these equalities, we obtain $\sum_{p-1}^{k-1} a_{p-k} = p^2(p - 1) - \sum_{k=1}^{p-1} a_k$, which yields $\sum_{k=1}^{p-1} a_k = \frac{p^2(p-1)}{2}$.

Problem 1.6. If $k \in \mathbb{Z}_+$, then show that
$$\left[\sqrt{k^2 + 1} + \cdots + \sqrt{k^2 + 2k}\right] = 2k^2 + 2k,$$
where $[x]$ denotes the integer part, i.e., the largest integer smaller than x.

Solution 1.6. We have $\sqrt{k^2 + m} - k = \frac{m}{\sqrt{k^2+m}+k}$, and if $1 \leq m \leq 2k$, then $\frac{m}{2k+1} \leq \frac{m}{\sqrt{k^2+m}+k} \leq \frac{m}{2k}$. Summing up all terms, we obtain
$$k = \sum_{m=1}^{2k} \frac{m}{2k+1} \leq \left(\sum_{m=1}^{2k} \sqrt{k^2 + m}\right) - 2k^2 \leq \sum_{m=1}^{2k} \frac{m}{2k} = k + \frac{1}{2},$$
whence the claim.

Problem 1.7. If n is not a multiple of 5, then $P = x^{4n} + x^{3n} + x^{2n} + x^n + 1$ is divisible by $Q = x^4 + x^3 + x^2 + x + 1$.

Solution 1.7. If ξ is a fifth primitive root of unity, then

$$P(\xi) = \xi^{4n} + \xi^{3n} + \xi^{2n} + \xi^n + 1 = \frac{\xi^{5n} - 1}{\xi^n - 1} = 0.$$

Hence $P = (x - \xi)(x - \xi^2)(x - \xi^3)(x - \xi^4)R = Q \cdot R$, for some polynomial R.

Comments 3 *More generally, $x^{(m-1)n} + \cdots + 1$ is divisible by $x^{m-1} + \cdots + 1$ if $\gcd(m, n) = 1$.*

Problem 1.8. Let p be an odd prime and $k \in \mathbb{Z}_+$. Show that there exists a perfect square the last k digits of whose expansion in base p are 1.

Solution 1.8. Use induction on k. It is obvious for $k = 1$. Let us assume that $x^2 = 1 + p + \cdots + p^{k-1} + (ap^k + \cdots)$ and $\gcd(x, p) = 1$. Then we have

$$(x + cp^k)^2 = x^2 + 2cxp^k + c^2 p^{2k} = 1 + p + \cdots + p^{k-1} + (a + 2cx)p^k + \cdots .$$

If p is odd, then $\gcd(2x, p) = 1$. Moreover, the congruence $2xc + a \equiv 1 \pmod{p}$ has at least one solution $c \in \{0, 1, \ldots, p - 1\}$. If c_0 is a solution, then $(x + c_0 p^k)^2$ has its last $k + 1$ digits equal to 1.

Comments 4 *The result can be extended to higher exponents $r \geq 2$ and prime numbers p that do not divide r, or more generally to p such that $\gcd(p, r) = 1$. Moreover, the last digits can be arbitrarily prescribed.*

Problem 1.9. Any natural number greater than 6 can be written as a sum of two numbers that are relatively prime.

Solution 1.9. For odd $n \geq 7$, we can write $n = 2k + 1 = k + (k + 1)$, where $\gcd(k, k+1) = 1$. Consider now $n = 2m$, where $m \geq 3$. Recall Chebyshev's theorem that for any $m \geq 2$ there exists a prime number p such that $m < p < 2m$. Therefore, we have $0 < 2m - p < m$. Further, $\gcd(p, 2m - p) = 1$, since $2m > p > m$ and $\gcd(m, p) = 1$.

Problem 1.10. Prove that it is impossible to extract an infinite arithmetic progression from the sequence $S = \{1, 2^k, 3^k, \ldots, n^k, \ldots\}$, where $k \geq 2$.

Solution 1.10. Let r be the ratio of an arithmetic progression $A \subset S$. If $n > r$, then $(n + 1)^k - n^k \geq 2n + 1 > r$. Thus the difference between consecutive terms of S grows higher than r, and hence A is finite.

Comments 5 *The result is also true when we replace S with the sequence $a_{n+2} = pa_{n+1} + qa_n$, for $1 \leq p \leq q + 1$.*

Problem 1.11. Prove that b^{a-j+1} divides $C_{b^a}^j$ if $a, b \geq 2$, $j \leq a + 1$.

Solution 1.11. An easy induction on n shows that $b^n \geq n+1$, and hence $b^a \geq j$. We write the combinations (also called binomial coefficients) as

$$C_{b^a}^j = \frac{1}{j!} \prod_{i=0}^{j-1} (b^a - i).$$

Let $i = b^r m$, where $\gcd(m, b) = 1$, so that $b^a - i = b^r(b^{a-r} - m)$. This shows that i and $b^a - i$ are divisible by the same power of b. Therefore, the exponent of b in $C_{b^a}^j$ equals the result of subtracting from a the exponent of b in $j!$. The latter is at most $j - 1$ because $b^{j-1} \geq j$, and hence the claim.

Problem 1.12. Solve in integers the following equations: (1) $x^2 = y^2 + y^3$; (2) $x^2 + y^2 = z^2$.

Solution 1.12. 1. The plane curve determined by the first equation has the following form:

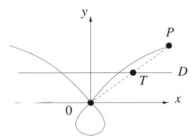

We will look for a parameterization of this curve by projecting from the origin 0 onto the line D given by $y = 1$. Thus, to each point P of the curve we associate the point T, which is the intersection of the half-line $0P$ with the line D. Denote by t the natural parameter of D. This amounts to setting $y = \frac{1}{t}x$ within the equation, which immediately yields $x = t^3 - t$ and $y = t^2 - 1$.

If $x, y \in \mathbb{Z}$, then $t \in \mathbb{Q}$. Moreover, $t^2 = y + 1 \in \mathbb{Z}$, but a rational number whose square is an integer must be an integer itself. Thus the integer solutions are given by the family $x = t^3 - t$ and $y = t^2 - 1$, where $t \in \mathbb{Z}$.

2. Dividing by z^2, we reduced the problem of finding the rational solutions of the equation $x^2 + y^2 = 1$. This is a circle of unit radius in the plane. We now use the stereographic projection from the point $P(-1, 0)$ onto the vertical line $x = 1$ in order to find a more-convenient parameterization of the circle.

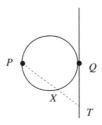

If t is the new parameter on the vertical line, then $x^2 + y^2 = 1$ and $\frac{x+1}{2} = \frac{y}{t}$. This yields $x = \frac{1-s^2}{1+s^2}$ and $y = \frac{2s}{1+s^2}$, where $s = \frac{t}{2}$.

Therefore the integer solutions of our equation are given by $x = (r^2 - s^2)v$, $y = 2vrs$, and $z = (r^2 + s^2)v$, where $s, r, v \in \mathbb{Z}$. The primitive solutions, for which $\gcd(x, y, z) = 1$, are those for which $v = 1$ and $\gcd(r, s) = 1$. The triples (x, y, z) are called Pythagorean triples.

Comments 6 *The two Diophantine equations above correspond to rational algebraic curves, i.e., to curves in the plane that admit a natural parameterization by means of rational functions. This situation is quite exceptional.*

Comments 7 *The Pythagorean equation $x^2 + y^2 = z^2$ can be generalized by adding more variables and still keeping its rational behavior to*

$$x_1^2 + x_2^2 + \cdots + x_{n+1}^2 = z^2, \quad n \geq 1.$$

The general (primitive) solution of this equation in integers can be obtained by the same method as above and reads

$$x_1 = s_1^2 + s_2^2 + \cdots + s_n^2 - u^2,$$
$$x_2 = 2s_1 u,$$
$$x_3 = 2s_2 u,$$
$$\vdots \quad \vdots$$
$$x_{n+1} = 2s_n u,$$
$$z = s_1^2 + s_2^2 + \cdots + s_n^2 + u^2,$$

for arbitrary integer parameters $s_1, s_2, \ldots, s_n, u \in \mathbb{Z}$.

Comments 8 *Another famous generalization of the Pythagorean equation is*

$$x^k + y^k = z^k, \quad k \geq 3,$$

which was conjectured by Fermat to have no nontrivial solutions for $k \geq 3$. Several particular cases were solved over the years ($k = 4$ by Fermat, $k = 3$ by Euler, $k = 5, 14$ by Dirichlet, $k = 7$ by Lamé, etc). However, it took more than three hundred years for the conjecture be eventually settled by A. Wiles in 1995.

Comments 9 *Euler considered the more general equation*

$$x_1^k + x_2^k + \cdots + x_n^k = z^k.$$

He conjectured that there are no nontrivial solutions unless $n \geq k$ and proved it for $n = 3$. However, Lander and Parkin found counterexamples for $n = 5$, for example,

$$27^5 + 84^5 + 110^5 + 133^5 = 144^5,$$

and later, N. Elkies settled the case n = 4 by finding another counterexample:

$$2\,682\,440^4 + 15\,365\,639^4 + 18\,796\,760^4 = 20\,615\,673^4.$$

A subsequent computer search by R. Frye found the following minimal solution for n = 4 (which is unique in the range z ≤ 1 000 000):

$$95\,800^4 + 217\,519^4 + 414\,560^4 = 422\,481^4.$$

See also

- N.D. Elkies: *On $A^4 + B^4 + C^4 = D^4$*, Math. Comp. 51 (1988), 184, 825–835.
- J.L. Lander and T.R. Parkin: *A counterexample to Euler's sum of powers conjecture.* Math. Comp. 21 (1967), 101–103.

Problem 1.13. Let $a, b, c, d \in \mathbb{Z}_+$ be such that at least one of a and c is not a perfect square and $\gcd(a, c) = 1$. Show that there exist infinitely many natural numbers n such that $an + b, cn + d$ are simultaneously perfect squares if one of the following conditions is satisfied:

1. b and d are perfect squares;
2. $a + b, c + d$ are perfect squares;
3. $a(d - 1) = c(b - 1)$.

Moreover, there do not exist such numbers if $a = 1, b = 0, c = 4k^2 - 1, d = 1$.

Solution 1.13. There exists n such that $an + b, cn + d$ are perfect squares iff the equation
$$ax^2 - cy^2 = ad - bc$$
has integer solutions. In fact, $\sqrt{cn + d}, \sqrt{an + b}$ are solutions of the equation. Conversely, if (x, y) is a solution, then
$$a(x^2 - d) = c(y^2 - b).$$

Since $\gcd(a, c) = 1$, we find that c divides $x^2 - d$, while a divides $y^2 - b$. In particular, we obtain that $x^2 - d = kc$ and $y^2 - b = ka$.

Further, ac is not a perfect square and so the Pell equation
$$u^2 - acv^2 = 1$$
has infinitely many solutions (u_n, v_n). If (x, y) is a solution of our equation, then
$$x_n = u_n x + dv_n y, \quad y_n = u_n y + v_n x$$
form an infinite sequence of solutions. Thus it suffices to determine one solution for the equation in order to find infinitely many.

The first two cases are immediate, and the third corresponds to the solution (1, 1).

Finally, note that the Pell-type equation $x^2 - (4k^2 - 1)y^2 = -1$ has no integer solutions, by modulo 4 considerations.

See also:

- T. Andreescu and D. Andrica, *Quadratic Diophantine Equations*, Springer Monographs in Math., 2006.

Problem 1.14. If $N = 2 + 2\sqrt{28n^2 + 1} \in \mathbb{Z}$ for a natural number n, then N is a perfect square.

Solution 1.14. If $\sqrt{28n^2 + 1} = k$, then $N = 2 + 2k$. If $N \in \mathbb{Z}$, then $k \in \frac{1}{2}\mathbb{Z}$. But k is the square root of an integer, and whenever such a square root is a rational number, then it is actually an integer. This proves that $k \in \mathbb{Z}$. Now, $28n^2 = (k-1)(k+1)$, and therefore k is odd, $k = 2t + 1$. Thus, $7n^2 = t(t+1)$. We have two cases:

1. If 7 divides $t + 1$, then, $t = 7s - 1$ and so $(7s-1)s = n^2$, for $s \in \mathbb{Z}$. But $\gcd(s, 7s-1) = 1$ and thus $7s - 1 = n_1^2$ and $s = n_2^2$, where $n_1 n_2 = n$. Moreover, any square, in particular n_1^2, is congruent to $1, 2$, or $4 \pmod 7$. In particular, it cannot be equal to $7s - 1$.

2. If 7 divides t, i.e., $t = 7s$, then $n^2 = s(7s+1)$. Since $\gcd(s, 7s+1) = 1$, we have $7s + 1 = n_1^2$ and $s = n_2^2$. This implies that

$$N = 2 + 2k = 2 + 2(14n_2^2 + 1) = 4(7n_2^2 + 1) = 4(7s + 1) = 4n_1^2.$$

Problem 1.15. Let $n \geq 5, 2 \leq b \leq n$. Prove that

$$\left[\frac{(n-1)!}{b}\right] \equiv 0 \pmod{b-1}.$$

Solution 1.15. We have four cases to consider:

1. If $b < n$, then $(n-1)!$ is divisible by $b(b-1)$ and therefore $\frac{(n-1)!}{b}$ is an integer divisible by $b - 1$.

2. Suppose $b = n$, where n is not prime and not the square of a prime number, hence $n = rs$, where $1 < r < s < n$. Since $\gcd(n, n-1) = 1$ and $s < n - 1$, we derive that $(n-1)!$ contains the factors r, s, and $n - 1$, and therefore it is divisible by $rs(n-1) = b(b-1)$.

3. If $b = n$ is the square of a prime, i.e., $n = p^2$, then $p \geq 3$ and hence $1 < p < 2p < 2p^2 - 1 = n - 1$. This implies that $(n-1)!$ is divisible by the product of $p, 2p$, and $n - 1$, i.e., by $2p^2(n-1) = 2b(b-1)$.

4. If $b = n$ is a prime number p, then by Wilson's theorem we have $(p-1)! + 1 \equiv 0 \pmod p$,

$$\left[\frac{(p-1)!}{p}\right] = \left[\frac{(p-1)!+1}{p} - \frac{1}{p}\right] = \frac{(p-1)!+1}{p} - 1 = (p-1)\left(\frac{(p-2)!-1}{p}\right).$$

Now $\gcd(p, p-1) = 1$, so $p - 1$ divides $\left[\frac{(p-1)!}{p}\right]$.

Problem 1.16. Prove that for every natural number n, there exists a natural number k such that k appears in exactly n nontrivial Pythagorean triples.

Solution 1.16. Let us show that 2^{n+1} appears in exactly n triples. If $n = 0$, it amounts to saying that 2 does not appear in any such triple, which is immediate.

Let us now use induction on n. All primitive Pythagorean triples are given by

$$x = u^2 - v^2, \quad y = 2uv, \quad z = u^2 + v^2,$$

where $\gcd(u, v) = 1$ and u, v are not both odd and $u > v$.

If u, v are both odd, then $y = 2uv = 2^{n+1}$ and therefore $u = 2^n$ and $v = 1$. The nonprimitive triples in which 2^{n+1} appears are those divisible by 2, and thus they are in bijection with the Pythagorean triples in which 2^n appears. By the recurrence hypothesis we have n such triples. Therefore 2^{n+1} appears in n nonprimitive triples and one primitive triple, and thus in $n + 1$ such triples.

Problem 1.17. Let $n, q \in \mathbb{Z}_+$ be such that all prime divisors of q are greater than n. Show that

$$(q - 1)(q^2 - 1) \cdots (q^{n-1} - 1) \equiv 0 \pmod{n!}.$$

Solution 1.17. Let $p \le n$, prime. Since p is not a divisor of q^k, we have $q^{k(p-1)} \equiv 1 \pmod{p}$, for all k. Therefore at least $\left[\frac{(n-1)}{(p-1)}\right]$ factors from the left-hand side are divisible by p.

Let us now compute the exponent of p in $n!$, which is

$$\exp_{n!}(p) = \sum_{k=1}^{\infty}\left[\frac{n}{p^k}\right] < \sum_{k}^{\infty}\frac{n}{p^k} = \frac{n}{p-1}.$$

Now $\exp_{n!}(p) \in \mathbb{Z}$, and so the strict inequality above implies that $\exp_{n!}(p) \le \left[\frac{n-1}{p-1}\right]$. Thus the left-hand side is divisible by $p^{\exp_{n!}(p)}$. Since this holds for all $p \le n$, the claim follows.

Comments 10 *Let $q = p^m$, where $p > n$ is a prime number. Let G denote the group of nonsingular matrices over the Galois field with q elements. Then the order of the group G is $|G| = q^{n(n-1)/2}(q-1) \cdots (q^n-1)$. Let now G^* be the subgroup generated by the diagonal matrices and their permutations. One finds that $|G^*| = (q-1)^n n!$. Since the order of a subgroup divides the order of the larger group, we derive*

$$(q - 1) \cdots (q^n - 1) \equiv 0 \pmod{(q^n - 1)^n n!},$$

improving the relation above.

Problem 1.18. Every natural number $n \ge 6$ can be written as a sum of distinct primes.

Solution 1.18. Let $n \ge 15$. One knows by Chebyshev's theorem that there exists a prime p between $\frac{n}{2}$ and n. Then $0 < n - p < \frac{n}{2}$. Further, there exists a prime q such that $\frac{n-p}{2} < q < n - p$, and the rest $n - p - q$ satisfies therefore $0 < n - p - q < \frac{n}{4}$. We continue this process until there remains either a prime number or a number from the set $\{7, 8, \ldots, 14\}$. Since $7 = 5 + 2$, $8 = 5 + 3$, $9 = 7 + 2$, $10 = 7 + 3$, $11 = 11$, $12 = 7 + 5$, $13 = 13$, and $14 = 2 + 5 + 7$, the claim follows.

Problem 1.19. Every number $n \geq 6$ can be written as a sum of three numbers which are pairwise relatively prime.

Solution 1.19. By Chebyshev's theorem , there exists a prime p such that $\frac{n}{2} < p < n$.

Note next that a natural number m such that m is divisible by all prime numbers strictly smaller than m must be 2. In fact, otherwise, again by Chebyshev's theorem, there exists a prime bigger than $\frac{m}{2}$, which cannot divide m. Thus there exists a prime q such that $n - p$ is not divisible by q, and so $\gcd(n - p, q) = 1$.

If $\gcd(p, n-p-q) > 1$, then $n - p - q = ap$, which cannot happen, since $p > \frac{n}{2}$ and $a \neq 0$, since $n - p - q > 0$. If $\gcd(q, n - p - q) > 1$, then $n - p - q = aq$, which contradicts our choice of q. Thus $p, q, n - p - q$ satisfy the claim.

Problem 1.20. Find all pairs of integers (m, n) such that

$$C_m^n = 1984,$$

where $C_m^n = \frac{m!}{n!(m-n)!}$ denotes the usual binomial coefficient.

Solution 1.20. We consider $0 \leq n \leq \frac{m}{2}$. We have $1984 = 2^6 \cdot 31$. Since $1984 = \frac{(n+1)(n+2)\cdots(m-1)m}{(m-n)!}$, then either 31 divides $(n + 1)(n + 2) \cdots (m - 1)m$ (and does not divide $(m - n)!$), so that $m \geq 31$, or else 31 also divides $(m - n)!$ and hence again $m \geq 31$.

If $n \geq 3$, then $C_m^n \geq C_m^3 \geq C_{31}^3 = \frac{1}{6}(29 \cdot 30 \cdot 31) > 64 \cdot 31 = 1984$. Thus $n \in \{0, 1, 2\}$.

Finally, if $n = 1$, then $m = 1984$, while $C_m^2 = 1984$ does not have any solutions in natural numbers.

Thus, the solutions are $(1984, 1)$, $(1984, 1983)$.

Comments 11 *More generally, if N has a prime divisor $p > k\left(\sqrt[k]{N} + 1\right)$, $k < \frac{p}{2}$, then the equation $C_m^n = N$ has at most $2k$ solutions. One derives, furthermore, that the number $F(N)$ of such solutions can be estimated from above:*

$$F(N) = \mathcal{O}(\log N / \log \log N).$$

Recall that $F(N) = \mathcal{O}(g(N))$ if $\lim_{N \to \infty} \frac{F(N)}{g(N)} < \infty$. It is conjectured that $F(N)$ is uniformly bounded, independently of N.

Problem 1.21. Find the set A consisting of natural numbers n that are divisible by all odd natural numbers a with $a^2 < n$.

Solution 1.21. Let $p_1 = 3, p_2, \ldots, p_m$ be the first m odd primes. Choose k such that $(2k - 1)^2 < n \leq (2k + 1)^2$ and m maximal such that $p_m \leq 2k - 1$. By hypothesis, $p_{m+1} > 2k + 1$ and hence, $n < p_{m+1}^2$.

Assume that n is in A. Then the p_i divide n for all $i \leq m$, and thus $p_1 p_2 \cdots p_m$ divides n. Moreover, the Bonse–Pósa inequality states that

$$p_1 \cdots p_m > p_{m+1}^2, \quad m \geq 4.$$

Thus $m \leq 3$.

If $m = 3$, then $n \leq 169$ and 105 divides n. However, $n = 105$ does not belong to A, since it is not divisible by 9.

If $m = 2$, then $n \leq 49$ and n is divisible by 15.

If $m = 1$, then $n \leq 25$ is divisible by 3.

If $m = 0$, then $n \leq 9$.

The answer is therefore $A = \{0, 1, 2, 3, 4, 5, 6, 7, 8, 9, 12, 15, 18, 21, 24, 30, 45\}$.

Comments 12 *Panaitopol recently gave an elementary proof of a more-general Bonse–Pósa inequality $p_1 p_2 \cdots p_n > p_{n+1}^{n-\pi(n)}$, for $n \geq 2$, which implies that $p_1 p_2 \cdots p_n > p_{n+1}^k$ for $n > 2k$. This has been further improved by Alzer and Berg.*

- L. Panaitopol: *An inequality involving prime numbers,* Univ. Beograd. Publ. Elektrotehn. Fak. Ser. Mat. 11 (2000), 33–35.
- H. Alzer, C. Berg: *Some classes of completely monotonic functions II,* The Ramanujan Journal 11(2006), 225–248.

Problem 1.22. Prove that $2^{1092} - 1$ is divisible by 1093^2.

Solution 1.22. Set $p = 1093$ and thus $p^2 = 1194649$. Observe that

$3^7 = 2187 = 2p + 1$,

$3^{14} \equiv 4p + 1 \pmod{p^2}$,

$2^{14} = 16348 = 15p - 11$, $2^{28} \equiv -330p + 121 \pmod{p^2}$,

$3^2 \cdot 2^{28} \equiv -2970p + 1089 \equiv -1876p - 4 \pmod{p^2}$, and therefore

$3^2 \cdot 2^{26} \equiv -469p - 1 \pmod{p^2}$. Therefore we infer that

$3^{14} \cdot 2^{182} \equiv -(489p + 1)^7 \equiv -3283p - 1 \equiv -4p - 1 \equiv -3^{14} \pmod{p^2}$,

and using the fact $\gcd(3^{14}, p^2) = 1$, we obtain $2^{182} \equiv -1 \pmod{p^2}$, yielding $2^{1092} - 1 \equiv 0 \pmod{1093^2}$.

Problem 1.23. Let $n \geq 0$, $r > 1$, and $0 < a \leq r$ be three integers. Prove that the number n, when written in base r, has precisely

$$\sum_{k=1}^{\infty} \left[nr^{-k} + ar^{-1} \right] - \left[nr^{-k} \right]$$

digits that are greater than or equal to $r - a$.

Solution 1.23. We write $n = n_1 + n_2 r + n_3 r^2 + \cdots$ in base r. Then the digits n_j are computed by the formulas

$$n_1 = n - \left[\frac{n}{r} \right] r, \quad n_2 = \left[\frac{n}{r} \right] - \left[\frac{n}{r^2} \right] r, \quad \ldots, \quad n_k = \left[\frac{n}{r^{k-1}} \right] - \left[\frac{n}{r^k} \right] r.$$

For a fixed k, we have $n_k \geq r - a$ if and only if $\left[\frac{n}{r^{k-1}} \right] - \left[\frac{n}{r^k} \right] r \geq r - a$, which is equivalent to asking that $\left[\frac{1}{r} \left(\left[\frac{n}{r^{k-1}} \right] + a \right) \right] \geq \left[\frac{n}{r^k} \right] + 1$.

Now let $n = mr^{k-1} + s$, where $0 \leq s < r^{k-1}$. Then

$$\left[\frac{1}{r}\left(\left[\frac{n}{r^{k-1}}\right]+a\right)\right]=\left[\frac{m}{r}+\frac{a}{r}\right]=\left[\frac{m}{r}+\frac{s}{r^k}+\frac{a}{r}\right]=\left[\frac{n}{r^k}+\frac{a}{r}\right]$$

and therefore $n_k \geq r - a$ if and only if $\left[\frac{n_1}{r^k}+\frac{a}{r}\right] > \left[\frac{n}{r^k}\right]$. However, we have

$$-\left[\frac{n}{r^k}\right]+\left[\frac{n}{r^k}+\frac{a}{r}\right]=\begin{cases} 0, & \text{if } n_k < r - a, \\ 1, & \text{if } n_k \geq r - a. \end{cases}$$

Thus, the total number of digits greater than $r - a$ is

$$\sum_{k=1}^{\infty}\left(\left[\frac{n}{r^k}+\frac{a}{r}\right]-\left[\frac{n}{r^k}\right]\right)$$

as claimed.

Comments 13 *The same argument shows that given a sequence $a_k \in \{1, 2, \ldots, r\}$, the number of digits n_k of n written in base r that satisfy the inequalities*

$$n_k \geq r - a_k$$

is

$$\sum_{k=1}^{\infty}\left(\left[\frac{n}{r^k}+\frac{a_k}{r}\right]-\left[\frac{n}{r^k}\right]\right).$$

Problem 1.24. Find a pair (a, b) of natural numbers satisfying the following properties:

1. $ab(a + b)$ is not divisible by 7;
2. $(a + b)^7 - a^7 - b^7$ is divisible by 7^7.

Solution 1.24. Since $a, b, a + b \not\equiv 0 \pmod 7$, there exists the inverse a^{-1} of a in $\mathbb{Z}/7^7\mathbb{Z}$; moreover, $k = ba^{-1} \not\equiv 0, -1 \pmod{7^7}$. Then 7^7 divides $(a^{-1})^7((a + b)^7 - a^7 - b^7) = (1 + k)^7 - 1 - k^7$. Developing the latter, we find that

$$7k(k^5 + 3k^4 + 5k^3 + 5k^2 + 3k + 1) \equiv 0 \pmod{7^7};$$

thus

$$(k + 1)(k^4 + 2k^3 + 3k^2 + 2k + 1) \equiv 0 \pmod{7^6}.$$

Since $k \not\equiv -1 \pmod{7^7}$, we obtain

$$(k^2 + k + 1)^2 \equiv 0 \pmod{7^6},$$

and so we have to solve in integers the equation $k^2 + k + 1 = 7^3 s$. The discriminant is $D = 4 \cdot 7^3 s - 3$, which has to be a square p^2. We need then to find p such that 7^3 divides $p^2 + 3$. We note that 7^2 divides $p^2 + 3$ only if either $p = 12 + 49q$ or $p = 37 + 49q$. By replacing above, we find in the first case $p^2 + 3 = 49(24q + 3) \equiv 0 \pmod{343}$, which yields $q \equiv -1 \pmod 7$ and thus $k = -19 + 343r$. The second case yields the solutions $k = 18 + 343r$.

In particular, $(1, 18)$ and $(1, 324)$ are solutions for the problem.

Problem 1.25. Let $0 < a < b < c < d$ be odd integers such that

1. $ad = bc$
2. $a + d = 2^k$, $b + c = 2^m$, for some integers k and m.

Prove that $a = 1$.

Solution 1.25. Let $\frac{a}{b} = \frac{c}{d} = \frac{x}{y}$, where $\gcd(x, y) = 1$. One obtains then, from the other equations, the value

$$a = \frac{x2^m(y - 2^{k-m}x)}{y^2 - x^2}.$$

Since a is odd, we have $y^2 - x^2 = 2^m s$ for some integer s. Recall that x, y are odd, since they divide a and b, respectively. We obtain, then, the following types of solutions:

1. $\begin{cases} y = 2^{m-2}s_1 + s_2, \\ x = 2^{m-2}s_1 - s_2, \end{cases}$

2. $\begin{cases} y = s_1 + 2^{m-2}s_2, \\ x = s_1 - 2^{m-2}s_2, \end{cases}$

where the s_i are both integers. Moreover, when replacing x and y by these values in the expression of a, we obtain that s_1 must divide $2^{k-m} + 1$ and s_2 must divide $2^{k-m} - 1$. Further, we have the equivalences

$$k > m \text{ iff } a + d > b + c \text{ iff } a + \frac{bc}{a} > b + c \text{ iff } (a - b)(a - c) > 0.$$

The inequality $a + \frac{bc}{a} \leq 1 + bc$ implies that $k < 2m - 2$.

Now, in the first type of solutions, we have $s_1 > 2^{m-2}s_2$, and s_1 divides $2^{k-m} + 1$, hence $s_2 = 1$, $s_1 = 2^{k-m} + 1$, $k = 2m - 2$.

In the second case, we have $y \leq b < 2^{m-1}$ and thus $s_1 < 2$, which yields $s_1 = 1$. Since $3y > 2^{m-1}$, we find that $x = a$, $y = b$, and replacing in the expression of a, we obtain $s_2 = 2^{m-2} - 1$, so that $a = 1$.

Problem 1.26. Find those subsets $S \subset \mathbb{Z}_+$ such that all but finitely many sums of elements from S (possibly with repetitions) are composite numbers.

Solution 1.26. Let $S^* = \{x = \sum_{a_i \in S} a_i\}$ be the semigroup generated by S. We prove that S^* contains only composite numbers iff there exists a prime number p, with $p \notin S$, such that p divides all elements of S. This condition is obviously sufficient. In order to prove the converse, let us assume the contrary, i.e., that $\gcd(S) = 1$. Then there exists a finite subset $T = \{c_1, \ldots, c_n\} \subset S$ such that $\gcd(T) = 1$. This implies that there exist integers $x_i \in \mathbb{Z}$ such that

$$\sum_{i=1}^{n} c_i x_i = 1.$$

Separating the terms with $x_i > 0$ and those with $x_i < 0$, we obtain two elements $a, b \in S^*$ with the property that $a - b = 1$. In particular, $\gcd(a, b) = 1$. From Dirichlet's theorem, the arithmetic progression $\{a + kb\}_{k \in \mathbb{Z}_+}$ contains infinitely many primes; but $a + kb \in S^*$, contradicting our assumptions. The claim follows.

Comments 14 *Another proof can be given by using the Chinese remainder theorem and showing that if S is such that $\gcd(S) = 1$, then S^* is $\mathbb{Z}_+ - A$, where A is a finite subset. The problem of determining the maximal element of A is called the Frobenius coin problem. If $S = \{a_1, a_2\}$, then the maximal number is $F(S) = a_1 a_2 - a_1 - a_2$, as was proved by Sylvester. If $\mathrm{card}(S) = 3$, the problem was solved by Selmer and Beyer, but for $\mathrm{card}(S) \geq 4$, there is no closed formula for $F(S)$. Erdős and Graham proved that*

$$F(\{a_1 < a_2 < \cdots < a_n\}) \leq 2a_{n-1} \left[\frac{a_n}{n} \right] - a_n,$$

and Brauer that

$$F(\{a_1 < a_2 < \cdots < a_n\}) \leq \sum_{i=1}^{n} a_i \left(\frac{\gcd(a_1, a_2, \ldots, a_i)}{\gcd(a_1, a_2, \ldots, a_{i+1})} - 1 \right).$$

Curtis proved that there is no closed formula expressing $F(S)$ for $n \geq 4$ that involves only polynomials.

- A. Brauer: *On a problem of partitions*, Amer. J. Math. 64 (1942), 299–312.
- F. Curtis: *On formulas for the Frobenius number of a numerical semi-group*, Math. Scand. 67 (1990), 2, 190–192.
- P. Erdős and R.L. Graham: *On a linear Diophantine problem of Frobenius*, Acta Arith. 21 (1972), 399–408.
- E.S. Selmer and Ö. Beyer: *On the linear Diophantine problem of Frobenius in three variables*, J. Reine Angew. Math. 301 (1978), 161–170.

Problem 1.27. Prove that for any natural number $n > 1$, the number $2^n - 1$ does not divide $3^n - 1$.

Solution 1.27. Set $A_n = 2^n - 1$ and $B_n = 3^n - 1$, where $n > 1$. For even n, 3 divides A_n, while 3 does not divide B_n, and so A_n does not divide B_n.

Now let $n > 1$ be odd, of the form $n = 2m - 1$. Using $2^4 \equiv 2^2 \pmod{12}$, we infer $A_n \equiv -5 \pmod{12}$. Since any prime greater than three is congruent to either ± 5, or ± 1 modulo 12, there is at least one prime divisor p of A_n such that $p \equiv \pm 5 \pmod{12}$. If A_n divides B_n, then p divides $3B_n$ and thus $3 \equiv 3^{n+1} \equiv 3^{2m} \pmod{p}$, which implies that 3 is a quadratic residue modulo p. But this contradicts the quadratic reciprocity theorem (an easy consequence of which is the fact that 3 cannot be a quadratic residue modulo p), since $p \equiv \pm 5 \pmod{12}$.

Problem 1.28. Let u_n be the least common multiple of the first n terms of a strictly increasing sequence of positive integers. Prove that

$$\sum_{n=1}^{\infty} \frac{1}{u_n} \leq 2.$$

Find a sequence for which equality holds above.

Solution 1.28. Let a_k be the increasing sequence, and $u_k = \text{lcm}(a_1, a_2, \ldots, a_k)$. We have

$$\sum_{k=1}^{\infty} \frac{1}{u_k} = \sum_{k=1}^{\infty} \frac{1}{\text{lcm}(a_1, a_2, \ldots, a_k)} \leq \frac{1}{a_1} + \sum_{k=2}^{\infty} \frac{1}{\text{lcm}(a_k, a_{k+1})}$$

$$= \frac{1}{a_1} + \sum_{k=2}^{\infty} \frac{\gcd(a_k, a_{k+1})}{a_k a_{k+1}} \leq \frac{1}{a_1} + \sum_{k=2}^{\infty} \frac{a_{k+1} - a_k}{a_k a_{k+1}}$$

$$= \frac{1}{a_1} + \sum_{k=2}^{\infty} \left(\frac{1}{a_k} - \frac{1}{a_{k+1}} \right) = \frac{2}{a_1} \leq 2.$$

Examples of sequences for which equality holds are $a_k = 2^{k-1}$, and $a_1 = 1, a_2 = 2, a_3 = 6, a_k = 3 \cdot 2^{k-2}$, for $k \geq 4$.

Problem 1.29. Let $\varphi_n(m) = \varphi(\varphi_{n-1}(m))$, where $\varphi_1(m) = \varphi(m)$ is the Euler totient function, and set $\omega(m)$ the smallest number n such that $\varphi_n(m) = 1$. If $m < 2^\alpha$, then prove that $\omega(m) \leq \alpha$.

Solution 1.29. If we consider the minimal j with $\varphi_j(m) = 1$, then $\varphi_{j-1}(m) = 2$. Also, we have $\varphi_1(m) < m < 2^\alpha$, but $\varphi_1(m)$ is an even number and thus $\varphi_2(m) \leq \frac{1}{2}\varphi_1(m) < 2^{\alpha-1}$. By induction, we obtain $\varphi_{j-1}(m) < 2^{\alpha-(j-2)}$, whence the claim. For even m, we have $\omega(m) \leq \alpha - 1$.

Problem 1.30. Let f be a polynomial with integer coefficients and $N(f) = \text{card}\{k \in \mathbb{Z}; f(k) = \pm 1\}$. Prove that $N(f) \leq 2 + \deg f$, where $\deg f$ denotes the degree of f.

Solution 1.30. Set $N_+(f) = \text{card}\{k; k \in Z, f(k) = 1\}$ and, respectively, $N_-(f) = \text{card}\{k; k \in Z, f(k) = -1\}$. We claim first that either $N_+(f) \leq 2$ or $N_-(f) \leq 2$. Assume the contrary. Then there exist a, b, c, d, e, g distinct integers such that

$$f(a) = f(b) = f(c) = -1 \text{ and } f(d) = f(e) = f(g) = +1.$$

Thus we can write

$$f(x) = Q(x)(x - a)(x - b)(x - c) - 1 = R(x)(x - d)(x - e)(x - g) + 1,$$

where Q and R are polynomials with integer coefficients. Consequently, we have the polynomial identity

$$(x - a)(x - b)(x - c)Q(x) = (x - d)(x - e)(x - g)R(x) + 2.$$

Let x take the values a, b, and c, respectively. Then we find that $a - d, a - e, a - g$ divide 2 and therefore $a - d, a - e, a - g \in \{1, -1, 2, -2\}$. In a similar way, $b - d, b - e, b - g \in \{1, -1, 2, -2\}$ and $c - d, c - e, c - g \in \{1, -1, 2, -2\}$. We can assume, from symmetry, that $a < b < c$ and $g < e < d$. We have then the strict inequalities $a - d < b - d < e - d < c - e < c - g$, and hence at least five distinct difference numbers that belong to $\{1, -1, 2, -2\}$, which is absurd. Thus our starting assumption is false.

In particular, if $N_+(f) \leq 2$, then $N(f) \leq N(f) + 2 \leq \deg f + 2$.

Comments 15 *The result can be improved as follows:*

$$N(f) - \deg f = \begin{cases} 1, & \text{if } \deg f = 1, \\ \leq 2, & \text{if } \deg f \in \{2, 3\}, \\ \leq 1, & \text{if } \deg f = 4, \\ \leq 0, & \text{if } \deg f \geq 5. \end{cases}$$

The solution is analogous. The equality for degree 2 can be attained, for instance, by taking $f(x) = x^2 - (2k+3)x + k^2 + 3k + 1$.

Problem 1.31. Prove that every integer can be written as a sum of 5 perfect cubes.

Solution 1.31. We have

$$6n = 0^3 + (n+1)^3 + (n-1)^3 + (-n)^3 + (-n)^3,$$

and this is written as a sum of four perfect cubes. Thus

1. $6n + 1 = 6n + 1^3$,
2. $6n + 2 = 6(n-1) + 2^3$,
3. $6n + 3 = 6(n-4) + 3^3$,
4. $6n + 4 = 6(n+2) + (-2)^3$,
5. $6n + 5 = 6(n+1) + (-1)^3$.

Comments 16 *Observe that we did not require the cubes to be positive. It is still unknown whether every integer is the sum of four cubes, although three cubes do not suffice. For instance, numbers that are congruent to ± 4 modulo 9 cannot be written as sums of three cubes, and moreover, it is not yet known whether they can be written as sums of four cubes. If we are looking for positive cubes, then the minimal number needed is 9, this being a particular case of the Waring problem. Notice that any sufficiently large number is a sum of seven positive cubes. The proof is more involved; for instance, see:*

- G.L. Watson: *A proof of the seven cube theorem.* J. London Math. Soc. 26 (1951), 153–156.

Problem 1.32. If $n \in \mathbb{Z}$, then the binomial coefficient $C_{2n}^n = \frac{(2n)!}{(n!)^2}$ has an even number of divisors.

Solution 1.32. The number of divisors of x, usually denoted by $\tau(x)$, is odd if and only if x is a perfect square. In fact, once we have a divisor d, there is a complementary divisor associated with it, namely $\frac{x}{d}$. These divisors are distinct unless x is a square and the divisor is its square root.

Further, it is known that for every natural number n, there exists a prime p satisfying $n < p \leq 2n$, and therefore $2n < 2p \leq 4n$. Writing $C_{2n}^n = \frac{(2n)!}{n!n!}$, we find that the prime p appears with exponent 1 in the development of C_{2n}^n and thus it cannot be a perfect square.

Problem 1.33. Prove that every $n \in \mathbb{Z}_+$ can be written in precisely $k(n)$ different ways as a sum of consecutive integers, where $k(n)$ is the number of odd divisors of n greater than 1.

Solution 1.33. We have $x + (x + 1) + (x + 2) + \cdots + (x + y) = \frac{1}{2}(y + 1)(2x + y)$. Therefore, we must find the number of natural solutions (x, y) of the Diophantine equation $\frac{1}{2}(y + 1)(2x + y) = n$. We have two cases to consider:

1. If $y = 2z$ is even, then $(2z + 1)(x + z) = n$. If we let $u = 2z + 1$, then u is an odd divisor of n. There is also associated the complementary divisor v, so that $uv = n$. Now $v = x + z$, and thus we find that $z = \frac{1}{2}(u - 1)$ and $x = \frac{1}{2}(2v - u + 1)$. Therefore (x, z) is an acceptable solution if and only if $2v - u \geq 1$.

2. If $y = 2z - 1$ is odd, then $z(2x + 2z - 1) = n$. Let $u = 2x + 2z - 1$, which is an odd divisor of S, and let v be the complementary divisor: $uv = n$. Then $z = v$; hence $x = \frac{1}{2}(u - 2v + 1)$. This implies that (x, z) is an acceptable solution if and only if $2v - u \leq -1$.

Now, given $u > 1$, an odd divisor of n, then either the first inequality or the second (but not both) could hold. Therefore, the number of solutions is that claimed.

Problem 1.34. Let $\pi_2(x)$ denote the number of twin primes p with $p \leq x$. Recall that p is a twin prime if both p and $p + 2$ are prime. Show that

$$\pi_2(x) = 2 + \sum_{7 \leq n \leq x} \sin\left(\frac{\pi}{2}(n + 2)\left[\frac{n!}{n + 2}\right]\right) \sin\left(\frac{\pi}{2}n\left[\frac{(n - 2)!}{n}\right]\right)$$

for $x > 7$.

Solution 1.34. If $n > 5$ is composite, then $\frac{(n-2)!}{n}$ is an even integer, and therefore $\sin\left(\frac{\pi}{2}n\left[\frac{(n-2)!}{n}\right]\right) = 0$.

If p is prime, then by Wilson's theorem, $(p - 2)! \equiv -(p - 1)! \equiv 1 \pmod{p}$, which implies that

$$\left[\frac{(p - 2)!}{p}\right] = \frac{(p - 2)! - 1}{p}.$$

If $p > 5$, then 4 divides $(p - 2)!$, and therefore

$$\sin\left(\frac{\pi}{2}p\left[\frac{(p - 2)!}{p}\right]\right) = \sin\left(\frac{\pi}{2}[(p - 2)! - 1]\right) = -1.$$

Consequently, the nth term of the sum is zero if at least one among $n, n + 2$ is composite, and 1 if both n and $n + 2$ are prime. Finally, one adds 2 units in order to count the pairs $(3, 5), (5, 7)$.

Problem 1.35.   1. Find all solutions of the equation $3^{x+1} + 100 = 7^{x-1}$.

2. Find two solutions of the equation $3^x + 3^{x^2} = 2^x + 4^{x^2}$, and prove that there are no others.

Solution 1.35. 1. $x = 4$ is a solution. If $x < 4$, then $3^{x+1} + 100 > 7^{x-1}$, and if $x > 4$, then $3^{x+1} + 100 < 7^{x-1}$. Therefore, this solution is unique.

2. Let $\rho(x) = \frac{4^{x^2} - 3^{x^2}}{3^x - 2^x}$. We have

$$\rho'(x) = \frac{(2 \cdot 4^{x^2} \ln 4 - 2 \cdot 3^{x^2} \ln 3) x (3^x - 2^x) - (4^{x^2} - 3^{x^2})(3^x \ln 3 - 2^x \log 2)}{(3^x - 2^x)^2}.$$

Therefore $\rho'(x) \geq (4^{x^2} - 3^{x^2})(3^x - 2^x)(x-1) \ln 3 = \mu(x)$. Next, we have $\mu(x) = 0$ if and only if $x \in \{0, 1\}$. These values are isolated roots of ρ', and thus ρ is strictly increasing for $x \in \mathbb{R}_+$. Therefore, the only two solutions of the equation are $x \in \{0, 1\}$, because, for other values of x, we have $\rho(x) > 1$.

Problem 1.36. Let $\sigma(n)$ denote the sum of the divisors of n. Prove that there exist infinitely many integers n such that $\sigma(n) > 2n$, or, even stronger, such that $\sigma(n) > 3n$. Prove also that $\sigma(n) < n(1 + \log n)$.

Solution 1.36. 1. Recall that if $n = p_1^{a_1} \ldots p_k^{a_k}$ is the factorization of n into prime factors, then we have the formula $\sigma(n) = \prod_{i=1}^{k} \frac{p_i^{a_i+1} - 1}{p_i - 1}$. Observe next that $n = 2^p 3^q$ satisfies the inequality $\sigma(n) > 2n$ for $p, q > 2$. Moreover, if $n = 2^p 3^q 5^s$, then $\sigma(n) > 3n$.

2. We have $\sigma(n) = \sum_{d/n} d$; hence

$$\frac{\sigma(n)}{n} = \sum_{d/n} \frac{d}{n} = \sum_{d/n} \frac{1}{d} = \sum_{d/n} \frac{1}{d} < 1 + \frac{1}{2} + \frac{1}{3} + \cdots + \frac{1}{n} < 1 + \log n.$$

Therefore $\sigma(n) < n(1 + \log n)$.

Problem 1.37. Let a_i be natural numbers such that $\gcd(a_i, a_j) = 1$, and a_i are not prime numbers. Show that:

$$\frac{1}{a_1} + \cdots + \frac{1}{a_n} < 2$$

Solution 1.37. Each a_i has at least one proper prime divisor p_i. Thus one can write $a_i = p_i q_i s_i$, where p_i and q_i are, not necessarily distinct, primes. Now, $\gcd(a_i, a_j) = 1$ implies that $p_i, q_i \neq p_j, q_j$. Therefore, the n numbers $\min(p_i, q_i)$ are distinct. We derive that

$$\sum \frac{1}{a_i} \leq \sum \frac{1}{p_i q_i} \leq \sum \frac{1}{(\min(p_i, q_j))^2} \leq \sum_{k \geq 1} \frac{1}{k^2} = \zeta(2) = \frac{\pi^2}{6} < 2.$$

Comments 17 *We denoted above by ζ the Riemann zeta function, defined by*

$$\zeta(s) = \sum_{k \geq 1} k^{-s}.$$

One knows the values of the zeta function at specific points. Let B_n denote the nth Bernoulli number, determined by the Taylor expansion

$$\frac{x}{e^x - 1} = \sum_{n=0}^{\infty} \frac{B_n}{n!} x^n.$$

The Bernoulli numbers are rational for all n and have the property that $B_{2n+1} = 0$ if $n \geq 1$. Their first values are $B_0 = 1$, $B_1 = -\frac{1}{2}$, $B_2 = \frac{1}{6}$, $B_4 = -\frac{1}{30}$, $B_6 = \frac{1}{42}$. Using contour integration, one finds the following expression for the Riemann zeta function in terms of Bernoulli numbers:

$$\zeta(2n) = (-1)^{n-1} \frac{(2\pi)^{2n}}{2(2n)!} B_{2n}$$

for all natural n. In particular, $\zeta(n)$ is transcendental for even n. It is considerably more difficult to show the irrationality at odd values. The arithmetic nature of the zeta values at odd positive integers has remained a mystery since Euler's time even though one now conjectures that the numbers π, $\zeta(3)$, $\zeta(5)$, ... are algebraically independent over the rationals. Apéry finally proved in 1979 that $\zeta(3)$ is irrational. Recent progress was made by Rivoal in 2000, who showed that there are infinitely many integers n such that $\zeta(2n + 1)$ is irrational. This result was subsequently tightened by Zudilin, who showed that one of $\zeta(5)$, $\zeta(7)$, $\zeta(9)$, and $\zeta(11)$ is irrational.

The Riemann zeta function has an analytic continuation to the entire complex plane except for a simple pole at $s = 1$. We have another marvelous identity concerning its values at negative integers and Bernoulli numbers:

$$B_n = (-1)^n n\zeta(1 - n)$$

for natural n.

- T. Rivoal: *La fonction zêta de Riemann prend une infinité de valeurs irrationnelles aux entiers impairs,* C. R. Acad. Sci. Paris Sér. I Math. 331 (2000), 4, 267–270.
- V.V. Zudilin: *One of the numbers $\zeta(5)$, $\zeta(7)$, $\zeta(9)$, $\zeta(11)$ is irrational,* Uspekhi Mat. Nauk 56 (2001), 4 (340), 149–150; translation in Russian Math. Surveys 56 (2001), 4, 774–776.

Problem 1.38. Let $\sigma(n)$ denote the sum of divisors of n. Show that $\sigma(n) = 2^k$ if and only if n is a product of Mersenne primes, i.e., primes of the form $2^k - 1$, for $k \in \mathbb{Z}_+$.

Solution 1.38. If $n = \prod_i p_i^{a_i}$, then $\sigma(n) = \prod_i \sigma(p_i^{a_i}) = \prod_i (1 + p_i + \cdots + p_i^{a_i})$. Moreover, if each prime factor is a Mersenne number, then $p_i = 2^{b_i} - 1$ and $a_i = 1$, which implies that $\sigma(n) = \prod_i 2^{b_i} = 2^k$.

Conversely, each prime factor p of n must satisfy $1 + \cdots + p^a = 2^s$ for some $s \in \mathbb{Z}$. This implies that $a = 2q + 1$ for $q \in \mathbb{Z}$ and furthermore, $2^s = (1 + p)(1 + p^2 + p^4 + \cdots + p^{2q})$. In particular, $1 + p = 2^e$ (thus p is a Mersenne prime) and $1 + p^2 + \cdots + (p^2)^q = 2^n$.

If $q > 0$, then the last equation implies again that q is odd, i.e., $q = 2v + 1$. We obtain next $(1 + p^2)(1 + p^4 + \cdots + p^{4v}) = 2^n$, and so $1 + p^2 = 2^w$. In particular, $1 + p = 2^e$ divides $2^w = 1 + p^2$. But $1 + p$ divides $1 - p^2$ and so $1 + p^2 + 1 - p^2 = 2$.

Since $1+p \geq 1+2 = 3$, we obtain a contradiction. Therefore, $q = 0$ and the exponent of the prime is $a = 1$, as claimed.

Problem 1.39. Find all integer solutions of the equation $|p^r - q^s| = 1$, where p, q are primes and $r, s \in \mathbb{Z} \setminus \{0, 1\}$.

Solution 1.39. The only solutions are $p = 3, q = 2, r = 2, s = 3$ and $p = 2, q = 3, r = 3, s = 2$. Obviously, one prime number should be even, hence 2. Let us assume that $q = 2$ and p is odd. Then the equation reads

$$p^r \pm 1 = 2^s.$$

If $r > 1$ is odd, then:
1. $(p^r + 1)/(p + 1)$ is the odd integer $p^{r-1} - p^{r-2} - p^{r-3} - \cdots - 1 > 1$, contradiction.
2. $(p^r - 1)/(p - 1)$ is the odd integer $p^{r-1} + p^{r-2} + p^{r-3} + \cdots + 1 > 1$, contradiction again.

Therefore $r = 2t$. Further, we have two cases for the equation above:
1. $2^s = \left(p^t\right)^2 + 1 = (2n + 1)^2 + 1 = 4n^2 + 4n + 2$. This is impossible since 4 divides 2^s since $s > 1$, while 4 does not divide the right-hand side $4n^2 + 4n + 2$.
2. $2^s = (p^t)^2 - 1 = 4n^2 + 4n = 4n(n + 1)$. Now, either n or $n + 1$ is odd, which would lead us to a contradiction as soon as $n \geq 2$. Thus $n = 1$ and the solutions are those claimed above.

Comments 18 *Catalan conjectured in 1844 that $3^2 - 2^3 = 1$ is the only solution of the Diophantine equation $x^m - y^n = 1$, in the integer unknowns $x, y, m, n \geq 2$. This was confirmed in 2002 by P. Mihăilescu, who gave a brilliant solution using cyclotomic fields. Note that it suffices to consider the case in which the exponents m, n are prime numbers. About one hundred years before Catalan, Euler had proved that there are no other solutions to the equation $|x^3 - y^2| = 1$, using the descent method. V.A. Lebesgue showed in 1850 that the equation $x^m - y^2 = 1$ has no solutions. More than one century later, in 1961, Chao Ko showed that the equation $x^2 - y^n = 1$ has no solutions if $n \geq 3$. His proof was improved by E. Z.Chein in 1976. More details concerning the history of the conjecture and various partial solutions can be found in the book of Ribenboim.*

- T. Metsänkylä: *Catalan's conjecture: another old Diophantine problem solved*, Bull. Amer. Math. Soc. (N.S.) 41(2004), 1, 43–57.
- P. Mihăilescu: *Primary cyclotomic units and a proof of Catalan's conjecture*, J. Reine Angew. Math. 572 (2004), 167–195.
- P. Ribenboim: *Catalan's Conjecture. Are 8 and 9 the Only Consecutive Powers?* Academic Press, Inc., Boston, MA, 1994.

Problem 1.40. Consider an arithmetic progression with ratio between 1 and 2000. Show that the progression does not contain more than 10 consecutive primes.

Solution 1.40. Suppose that there are 11 consecutive prime terms in the progression. Let r be the ratio. The respective terms are:

$$h + nr, h + nr + r, h + nr + 2r, \ldots, h + nr + 10r.$$

Let p be a prime number less than 12. If $r \not\equiv 0 \pmod{p}$, then there exists some $i \leq 10$ such that $h + ir \equiv 0 \pmod{p}$, and hence $h + ir$ will not be prime. This shows that r is divisible by 2, 3, 5, 7, 11 and thus greater than 2000, contradicting our assumptions.

Comments 19 *More generally, if the ratio r is less than $p_1 \cdots p_n$, then the maximum number of consecutive primes is at most $p_n - 1$.*

Problem 1.41. Let $a_1 = 1$, $a_{n+1} = a_n + \left[\sqrt{a_n}\right]$. Show that a_n is a perfect square iff n is of the form $2^k + k - 2$.

Solution 1.41. Let $n_k = 2^k + k - 2$. We will prove by induction first that $a_{n_k} = (2^{k-1})^2$ and second that $n_{k-1} < m < n_k$ implies that a_m is not a perfect square. The assention is true for $k = 1$. One proves by induction on i that the formulas

$$a_{n_k + 2i} = (2^{k-1} + 1 - i)^2 + 2k,$$

$$a_{n_k + 2i + 1} = (2^{k-1} + i)^2 + 2^{k-1} - i,$$

hold for $0 \leq i \leq 2^{k-1}$. In particular, if $i = 2^{k-1}$, we obtain $a_{n_{k+1}} = (2^k)^2$. The other values are not perfect squares, since they are located between two consecutive squares.

Problem 1.42. Recall that $\varphi(n)$ denotes the Euler totient function (i.e., the number of natural numbers less than n and prime to n), and that $\sigma(n)$ is the sum of divisors of n. Show that n is prime iff $\varphi(n)$ divides $n - 1$ and $n + 1$ divides $\sigma(n)$.

Solution 1.42. If n is prime, then $\varphi(n) = n - 1$, $\sigma(n) = n + 1$. Conversely, let us assume $n \geq 3$. Then $\varphi(n)$ is even and thus n must be odd. If p is an odd prime such that p^r divides n and $r \geq 2$, then p^{r-1} divides $\varphi(n)$ and thus p^{r-1} divides $n - 1$, which is a contradiction, since n and $n - 1$ are coprime, i.e., $\gcd(n, n - 1) = 1$. Therefore, we have $n = p_1 \cdots p_k$, where p_i are odd primes.

We have further, $\varphi(n) = (p_1 - 1) \cdots (p_k - 1)$ and $\sigma(n) = (p_1 + 1) \cdots (p_k + 1)$. This implies that 2^k divides both $\varphi(n)$ and $\sigma(n)$.

If $k \geq 2$, then 4 divides $\varphi(n)$ and hence $n - 1$. This implies that $\gcd(4, n + 1) = 2$. Since 2^k divides $\sigma(n)$, we find that 2^{k-1} must divide $\frac{\sigma(n)}{n+1}$. In particular, $2^{k-1} < \frac{\sigma(n)}{n}$, but

$$\frac{\sigma(n)}{n} = \left(1 + \frac{1}{p_1}\right) \cdots \left(1 + \frac{1}{p_k}\right) < \left(\frac{4}{3}\right)^k,$$

which leads to a contradiction when $k \geq 2$. Therefore, $k = 1$ and so n is prime.

Comments 20 *An old conjecture due to Lehmer states that if $\varphi(n)$ divides $n - 1$, then n is prime. P. Hagis Jr. proved that the number of prime divisors $\tau(n)$ is either 1 or $\tau(n) \geq 298\,848$ and $n > 10^{1937042}$ if 3 divides n, and $\tau(n) \geq 1991$ and $n > 10^{8171}$ in general. The reader may consult:*

- P. Hagis Jr.: *On the equation $M \cdot \varphi(n) = n - 1$*, Nieuw Arch. Wisk. (4) 6 (1988), 3, 255–261.

Problem 1.43. A number is called φ-subadditive if $\varphi(n) \leq \varphi(k) + \varphi(n-k)$ for all k such that $1 \leq k \leq n-1$, and φ-superadditive if the reverse inequality holds. Prove that there are infinitely many φ-subadditive numbers and infinitely many φ-superadditive numbers.

Solution 1.43. Let $p \geq 3$ prime and $1 < k < p$. We have then

$$\varphi(k) + \varphi(p-k) \leq k - 1 + (p-k-1) = p - 2 < \varphi(p).$$

Now $p - 1$ is composite; thus $\varphi(p-1) \leq p - 1$, and hence $\varphi(1) + \varphi(p-1) \leq 1 + p - 3 < \varphi(p)$. Therefore, every prime number p is φ-superadditive.

Let $r \geq 2$ and set $n_r = p_1 \cdots p_r$ for the product of the first r prime numbers. Consider some natural number m, such that $1 < m < n_r$. Let $q_1, \ldots, q_s$ be the prime divisors of m. We have $q_j \geq p_j$ and thus $s \leq r$, since otherwise we would have $m \geq q_1 \cdots q_s > p_1 \cdots p_r = n_r$. Further, one computes

$$\frac{\varphi(m)}{m} \geq \prod_{i=1}^{s}\left(1 - \frac{1}{q_i}\right) \geq \prod_{i=1}^{r}\left(1 - \frac{1}{p_i}\right) = \frac{\phi(n_r)}{n_r}.$$

This implies that

$$\varphi(m) + \varphi(n_r - m) \geq m\frac{\varphi(n_r)}{n_r} + (n_r - m)\frac{\varphi(n_r)}{n_r} = \varphi(n_r).$$

Problem 1.44. Find the positive integers N such that for all $n \geq N$, we have $\varphi(n) \leq \varphi(N)$.

Solution 1.44. 1. Let $N \geq 5$. According to Chebyshev's theorem, there exists a prime number p such that

$$p\left[\frac{N+3}{3}\right] < p < 2\left[\frac{N+3}{2}\right] - 2 \leq N + 2.$$

We have $\varphi(p) = p-1 > \frac{N+1}{2} \geq \varphi(N+1)$ for odd N and $\varphi(p) = p-1 > \frac{N}{2} \geq \varphi(N)$ for even N.

This means that $N \leq 5$, and since $\varphi(1) = \varphi(2) = 1$, $\varphi(3) = \varphi(4) = 2$, we have $N \in \{1, 2, 3, 4\}$.

Problem 1.45. A number n is perfect if $\sigma(n) = 2n$, where $\sigma(n)$ denotes the sum of all divisors of n. Prove that the even number n is perfect if and only if $n = 2^{p-1}(2^p - 1)$, where p is a prime number with the property that $2^p - 1$ is prime.

Solution 1.45. Since n is even, one can write $n = 2^q m$, where $q \geq 1$ and m is odd. One knows that $\sigma(p^a) = \frac{p^{a+1}-1}{p-1}$ if p is prime, and $\sigma(p_1^{a_1} p_2^{a_2} \cdots p_s^{a_s}) = \sigma(p_1^{a_1})\sigma(p_2^{a_2})\cdots\sigma(p_s^{a_s})$ if p_j are distinct primes.

We have then $\sigma(m) > m$, and we write $\sigma(m) = m + r$ for some natural number $r \geq 1$. The hypothesis $\sigma(n) = 2n$ is equivalent then to $2^{q+1}m = (2^{q+1}-1)(m+r) = 2^{q+1}m - m + (2^{q+1}-1)r$. Thus $m = (2^{q+1}-1)r$. We have found therefore that r divides m and $r < m$. But $\sigma(m) = m + r$ is the sum of all divisors of m, and thus r is the sum of all divisors of m that are strictly smaller than m. Since r itself is such a divisor, it follows that it cannot be any other divisor of m smaller than r and so $r = 1$. In particular, this means that m is prime.

This means that $\sigma(m) = m + 1 = 2^{q+1} - 1$ and $2^{q+1}m = (2^{q+1}-1)(m+1)$. Since m is odd and relatively prime to $m + 1$, it follows that $m = 2^{q+1} - 1$. Further, m has to be prime and so $p = q + 1$ is a prime number. In fact, if p is composite, say $p = ab$, then $2^p - 1 = (2^a - 1)(1 + 2^a + \cdots + 2^{a(b-1)})$. This proves the claim.

Comments 21 *Perfect numbers were known in antiquity. Euclid stated in Book IX of his* Elements *that a number of the form $2^{p-1}(2^p - 1)$ for which the odd factor is prime is a perfect number, and noticed that $6, 28, 496,$ and 8128 are perfect. Descartes claimed in 1638, in a letter to Mersenne, that every even perfect number should be of this form, and this was proved later by Euler to be true. Euler also considered the problem of the existence of odd perfect numbers. Although numbers up to 10^{300} have been checked, no odd perfect numbers have been found to this day, supporting the conjecture that there are no odd perfect numbers. This is one of the oldest open problems in mathematics, along with the congruent number problem.*

Dickson proved in 1913 that for each k, there are at most finitely many odd perfect numbers with k distinct prime factors, and the best estimate known (due to Pace P. Nielsen) is that such a number is smaller than 2^{4^k}. Here are some properties that an odd perfect number must satisfy:

- *P. Hagis Jr. and, independently, J.E.Z. Chein proved that it has at least 8 distinct prime factors (at least 11 if it is not divisible by 3);*
- *H.G. Hare showed that it has at least 47 prime factors in total, including repetitions;*
- *P. Jenkins proved that it has at least one prime factor greater than 10^7, and D.E. Iannucci showed that it has two prime factors greater than 10^4, and three prime factors greater than 100;*
- *it is divisible by the power of a prime number that is greater than 10^{20};*
- *it is a perfect square multiplied by an odd power of a single prime.*

Primes of the form $2^p - 1$ (so p is also prime) are called Mersenne primes. It is still unknown whether there exist infinitely many Mersenne primes. The first few such primes correspond to $p = 2, 3, 5, 7, 13, 17, 19, 31, 61, 89$. As of October 2006, only 44 Mersenne primes were known; the largest known prime number, $2^{32\,582\,657} - 1$, is a Mersenne prime having $9\,808\,358$ digits. It was discovered in September 2006 by Curtis Cooper and Steve Boone.

- R.K. Guy: *Unsolved Problems in Number Theory,* section B1, 2nd ed. New York, Springer-Verlag, 44–45, 1994.

- P. Hagis, Jr.: *An outline of a proof that every odd perfect number has at least eight prime factors,* Math. Comput. 34 (1980), 1027–1032.
- D.E. Iannucci: *The second largest prime divisor of an odd perfect number exceeds ten thousand,* Math. Comput. 68 (1999), 1749–1760.
- D.E. Iannucci: *The third largest prime divisor of an odd perfect number exceeds one hundred,* Math. Comput. 69 (2000), 867–879.
- P. Jenkins: *Odd perfect numbers have a factor that exceeds 10^7,* Math.Comput. 72 (2003), 1549–1554.
- P.P. Nielsen: *An upper bound for odd perfect numbers,* Integers 3(2003), A14, (electronic).

Problem 1.46. A number n is superperfect if $\sigma(\sigma(n)) = 2n$, where $\sigma(k)$ is the sum of all divisors of k. Prove that the even number n is superperfect if and only if $n = 2^r$, where r is an integer such that $2^{r+1} - 1$ is prime.

Solution 1.46. If $n = 2^r$ and $2^{r+1} - 1$ is prime, then $\sigma(\sigma(n)) = \sigma(2^{r+1} - 1) = 2^{r+1} = 2n$. Conversely, let us consider $n = 2^r q$, where q is odd and $r \geq 1$. Then we have the identity

$$2^{r+1}q = 2n = \sigma(\sigma(n)) = \sigma((2^{r+1} - 1)\sigma(q)).$$

Moreover, if $q > 1$, then the number $(2^{r+1} - 1)\sigma(q)$ has at least three distinct divisors, namely $(2^{r+1} - 1)\sigma(q)$, $\sigma(q)$, and $2^{r+1} - 1$. Thus

$$\sigma((2^{r+1} - 1)\sigma(q)) \geq 2^{r+1}\sigma(q) + 2^{r+1} - 1 > 2^{r+1}q,$$

which is a contradiction. Therefore $q = 1$ and $2^{r+1} = \sigma(2^{r+1} - 1)$; thus $2^{r+1} - 1$ is prime, as claimed.

Comments 22 *It is still unknown whether there exist odd superperfect numbers. If one exists, it must be a square and greater than $7 \cdot 10^{24}$, as was proved by J.L. Hunsucker and C. Pomerance. Earlier, in 1967, D. Suryanarayana proved that there do not exist odd superperfect numbers that are powers of a prime.*

- J.L. Hunsucker and C. Pomerance: *There are no odd super perfect numbers less than $7 \cdot 10^{24}$,* Indian J. Math. 17 (1975), 3, 107–120.

Problem 1.47. If a, b are rational numbers satisfying $\tan a\pi = b$, then $b \in \{-1, 0, 1\}$.

Solution 1.47. Write $a = \frac{m}{n}$, where $\gcd(m, n) = 1$. De Moivre's identities read

$$(\cos a\pi + i \sin a\pi)^n = \cos m\pi + i \sin m\pi, \quad (\cos a\pi - i \sin a\pi)^n = \cos m\pi - i \sin m\pi.$$

Therefore

$$\left(\frac{1 + ib}{1 - ib}\right)^n = \left(\frac{1 + i\tan a\pi}{1 - i\tan a\pi}\right)^n = \left(\frac{\cos a\pi + i \sin a\pi}{\cos a\pi - i \sin a\pi}\right)^n = 1.$$

Consider the polynomial $P = x^n - 1 \in \mathbb{Z}[i][x]$. Since P is a monic polynomial and $\mathbb{Z}[i]$ is a factorial ring, any root of P that lies in $\mathbb{Q}[i]$ actually belongs to the ring of integers $\mathbb{Z}[i]$. This means that

$$\frac{1 + ib}{1 - ib} = \frac{1 - b^2}{1 + b^2} + i\frac{2b}{1 + b^2} \in \mathbb{Z}[i]$$

and hence $\frac{1-b^2}{1+b^2} \in \mathbb{Z}$, $\frac{2b}{1+b^2} \in \mathbb{Z}$. Now it is immediate that $b \in \{-1, 0, 1\}$.

Problem 1.48. Let $A_n = ru^n + sv^n$, $n \in \mathbb{Z}_+$, where r, s, u, v are integers, $u \neq \pm v$, and P_n be the set of prime divisors of A_n. Then $P = \bigcup_{n=0}^{\infty} P_n$ is infinite.

Solution 1.48. We suppose that $rsuv \neq 0$. We can assume, without loss of generality, that $\gcd(ru, sv) = 1$, and so the prime numbers in P do not divide $rsuv$. Let us suppose that P is finite, say $\{p_1, p_2, \ldots, p_m\}$.

We have $u^i \neq v^i$, for any $i \neq 0$, because $u \neq \pm v$. Then there exists some a such that $u^i \not\equiv v^i \pmod{p_k^a}$ for all $1 \leq i \leq m+1$, $1 \leq k \leq m$. Moreover, if $1 \leq i < j \leq m+1$, then

$$A_j - v^{j-i}A_i = ru^j\left(u^{j-i} - v^{j-i}\right),$$

and therefore we cannot have $A_i \equiv A_j \pmod{p_k^a}$, for all $k \leq m$. Thus there exists t within the range $1 \leq t \leq m+1$ such that $A_t \not\equiv 0 \pmod{p_k^a}$ for $1 \leq k \leq m$. We choose now some b such that $u^b \equiv v^b \pmod{p_k^a}$ for all $1 \leq k \leq m$. We can take, for instance, $b = \varphi(c^a)$, where φ is Euler's totient function and $c = p_1 \cdots p_m$. We now choose n large enough that $A_{t+nb} > c^a$. Since we have $A_{t+nb} \equiv A_t \not\equiv 0 \pmod{p_k^a}$ for all $1 \leq k \leq m$, we obtain that A_{t+nb} must also have a prime divisor different from $p_1, \ldots, p_m$.

Comments 23 *This problem is a particular case of a result of G. Pólya, see also:*

- G. Pólya: *Arithmetische Eigenschaften der Reihenentwiclungen rationalen Functionen*, J. Reine Angew. Math. 151 (1921), 1–31.

Problem 1.49. Solve in natural numbers the equation

$$x^2 + y^2 + z^2 = 2xyz.$$

Solution 1.49. If x, y, z are odd, then the left-hand side is odd and therefore nonzero. Thus $2xyz \equiv 0 \pmod 4$. Each square is congruent to either 0 or 1 modulo 4, and thus x, y, z must be even. Set $x = 2x_1, y = 2y_1, z = 2z_1$. The equation becomes $x_1^2 + y_1^2 + z_1^2 = 4x_1 y_1 z_1$. The same argument shows that x_1, y_1, z_1 are even. By induction, we find a sequence of triples with $x_n = 2x_{n+1}, y_n = 2y_{n+1}, z_n = 2z_{n+1}$, satisfying

$$x_{n+1}^2 + y_{n+1}^2 + z_{n+1}^2 = 2^{n+2}x_{n+1}y_{n+1}z_{n+1}.$$

Moreover, if $n \geq \max(|x|, |y|, |z|)$, then $|x_n| = |\frac{1}{2^{n+1}}x| < 1$ and similarly for the other unknowns. Since x_n is an integer, we obtain $x_n = 0$ and thus $x = y = z = 0$.

Problem 1.50. Find the greatest common divisor of the following numbers: C_{2n}^1, C_{2n}^3, C_{2n}^5, ..., C_{2n}^{2n-1}.

Solution 1.50. We have the identity $C_{2n}^1 + \cdots + C_{2n}^{2n-1} = 2^{2n-1}$, which can be obtained by writing the binomial expansion of $(x + 1)^n$ and evaluating it at $x = 1$ and $x = -1$. Therefore, the greatest common divisor g should be of the form 2^p, for some natural number p.

If $n = 2^k q$, where q is odd, then $C_{2n}^1 = 2^{k+1}q$ and hence $g \leq 2^{k+1}$. But now we can compute $C_{2n}^p = C_{2^{k+1}q}^p = \frac{2^{k+1}q}{p} C_{2^{k+1}q-1}^{p-1}$. Since p is odd and the binomial coefficient is an integer, it follows that 2^{k+1} divides C_{2n}^p for all odd p. This proves that the number we are seeking is $g = 2^{k+1}$.

Problem 1.51. Prove that if $S_k = \sum_{i=1}^n k^{\gcd(i,n)}$, then $S_k \equiv 0 \pmod{n}$.

Solution 1.51. 1. $\sum_{i=1}^n k^{\gcd(i,n)} = \sum_{d/n} \varphi\left(\frac{n}{d}\right) k^d$. If $\gcd(m, n) = 1$, then we have the following identity:

$$\sum_{i=1}^{mn} k^{\gcd(i,mn)} = \sum_{d/m} \varphi\left(\frac{m}{d}\right) \sum_{\tau/n} \varphi\left(\frac{n}{\tau}\right) (k^d)^\tau.$$

Thus it suffices to prove the claim for $n = p^\alpha$, where p is prime.

The case $\alpha = 1$ or p divides k is trivial. Let $\alpha > 1$ and $\gcd(p, k) = 1$. We will prove the claim by induction on α, as follows:

$$\sum \varphi\left(\frac{p^\alpha}{p^i}\right) k^{p^i} = k^{p^\alpha} + \varphi(p)k^{p^{\alpha-1}} + \varphi(p^2)k^{p^{\alpha-2}} + \cdots + \varphi(p^\alpha)k$$

$$\equiv k^{p^{\alpha-1}} + (p - 1)k^{p^{\alpha-1}} + \cdots + \varphi(p^\alpha)k$$

$$\equiv p(k^{p^{\alpha-1}} + \cdots + \varphi(p^{\alpha-1}))k$$

$$\equiv p \sum_{i=0}^{\alpha-1} \varphi\left(\frac{p^{\alpha-1}}{p^i}\right) k^{p^i} \pmod{p^\alpha}.$$

This proves the induction step.

2. One observes that $\frac{1}{n} S_k$ counts the number of circular permutations of length k on n elements. Hence, $S_k \equiv 0 \pmod{n}$.

Problem 1.52. Prove that for any integer n such that $4 \leq n \leq 1000$, the equation

$$\frac{1}{x} + \frac{1}{y} + \frac{1}{z} = \frac{4}{n}$$

has solutions in natural numbers.

Solution 1.52. It suffices to consider the case n prime, since a solution for n yields a solution for a multiple of n. Let $a, b, c, d \in \mathbb{Z}_+$. If

$$an + b + c = 4abcd,$$

then $(x, y, z) = (bcd, abdn, acdn)$ is a solution. In particular, take $a = 2, b = 1, c = 1$. Then, if $n = 4d - 1$, we obtain a solution as above. This works actually for any $n \equiv -1 \pmod 4$. Taking $a = 1, b = 1, c = 1$, we solve the case $n \equiv 2 \pmod 4$. If $n \equiv 0 \pmod 4$, then we have the solutions $x = y = z = 3n/4$. It remains to check the case $n \equiv 1 \pmod 4$.

Further, take $a = 1, b = 1, c = 2$, and thus we solve the case $n \equiv 3 \pmod 8$. The only case remaining is then $n \equiv 1 \pmod 8$.

If $n = 3$, we have the solutions $(x, y, z) = (3, 2, 2)$. Write

$$an + b = c(4abd - 1) = cq, \text{ where } q \equiv -1 \pmod{4ab}.$$

Take $q = 3, a = b = 1$; then $n \equiv -1 \pmod 3$ is solved. There remains the case $n \equiv 1 \pmod 3$, which we will suppose from now on.

If $n = 7$, then we have the solution $(x, y, z) = (2, 28, 28)$. Next take $q = 7$ and $ab \in \{1, 2\}$. This leads to solutions for $n \equiv -1, -2, -4 \pmod 7$. Thus it suffices to consider $n \equiv 1, 2, 4 \pmod 7$.

Take $q = 15$ and $a = 2, b = 1$ or $a = 1, b = 2$. This solves the equation for $n \equiv -2, -8 \pmod{15}$. Since $n \equiv 1 \pmod 3$, we have $n \equiv -2, -3 \pmod 5$. Hence, we may suppose that $n \equiv 1, 4 \pmod 5$.

By asking that n satisfy all these congruences, it follows that the problem is solved for all n such that

$$n \not\equiv 1, 121, 169, 289, 361, 529 \pmod{840}.$$

Since the first prime number not satisfying this condition is 1009, the claim follows.

Comments 24 *Erdős and Straus conjectured that the equation $\frac{1}{x} + \frac{1}{y} + \frac{1}{z} = \frac{4}{n}$ has solutions for every $n \geq 4$. Mordell proved that this conjecture is true, except possibly for the primes in six particular residue classes modulo 840. Vaughan showed that a sufficient condition for the more general conjecture in which $\frac{4}{n}$ is replaced by $\frac{m}{n}$ is the existence of positive integers a, b for which $an + b$ has a divisor congruent to $-1 \pmod{mab}$. The conjecture was verified by I. Kotsireas, via computer search, for $n \leq 10^{10}$.*

Schinzel conjectured further that the equation

$$\frac{m}{n} = \frac{1}{x_1} + \cdots + \frac{1}{x_k}$$

has positive integer solutions x_i if $m > k \geq 3$; actually, the truth of the conjecture for $m > k = 3$ implies its validity for every $m > k \geq 3$. See also the book of R.K. Guy which contains more partial results on this topic.

- P. Erdős: *On a Diophantine equation* (Hungarian, English summary) Matematikai Lapok 1 (1950), 192–210.
- R.K. Guy: *Unsolved Problems in Number Theory,* Section D11, Springer, New York, 1981.

- L.J. Mordell: *Diophantine Equations,* Pure and Applied Mathematics, Vol. 30 Academic Press, London-New York 1969.
- A. Schinzel: *Sur quelques propriétés des nombres 3/n et 4/n, où n est un nombre impair,* Mathesis 65 (1956), 219–222.
- W. Sierpiński: *Sur les décompositions de nombres rationnels en fractions primaires,* Mathesis 65 (1956), 16–32.

Problem 1.53. For a natural number n, we set $S(n)$ for the set of integers that can be written in the form $1 + g + \cdots + g^{n-1}$ for some $g \in \mathbb{Z}_+, g \geq 2$.

1. Prove that $S(3) \cap S(4) = \emptyset$.
2. Find $S(3) \cap S(5)$.

Solution 1.53. 1. Let $g, h \geq 2$ be integers satisfying the equation

$$1 + g + g^2 = 1 + h + h^2 + h^3$$

and therefore $h^3 = (g - h)(g + h + 1)$. Any prime divisor p of h divides $g - h$ or $g + h + 1$.

In the first case, p divides g, and in the second case, p divides $g + 1$. Furthermore, if p^m divides h then either p^{3m} divides $g - h$ or p^{3m} divides $g + h + 1$, because $\gcd(g, g + 1) = 1$. Thus we can write $h = h_1 h_2$, where $\gcd(h_1, h_2) = 1$, so that

$$h_1^3 = g - h, \quad h_2^3 = g + h + 1.$$

In particular, we have $2h_1 h_2 = 2h = h_2^3 - h_1^3 - 1$.

We will show that this Diophantine equation does not have any solutions $h_1, h_2 \geq 0$. Let $f_{h_1}(h_2) = h_2^3 - h_1^3 - 2h_1 h_2 - 1$. First, we have $f_{h_1}(h_2) \leq -1$ if $0 \leq h_2 \leq h_1$. Moreover, if $h_2 > h_1$, then f_{h_1} is increasing and so f_{h_1} does not have any natural zeros, since

$$f_{h_1}(h_1 + 1) = h_1^2 + h_1 > 0.$$

This proves that $S(3) \cap S(4) = \emptyset$.

2. Let $f(x) = 1 + x + x^2$ and $g(x) = 1 + x + x^2 + x^3 + x^4$. Then f and g are strictly increasing functions on $\mathbb{Z}_+$. Moreover, we have the following inequalities:

$$[f(4n^2 + 5n + 1) < g(2n + 1) < f(4n^2 + 9n + 4) < g(2n + 2)$$
$$< f(4n^2 + 9n + 5) < f(4(n + 1)^2 + 5(n + 1) + 1), \text{ for all } n \geq 1.$$

This implies that $g(n) \neq f(k)$ if $n \geq 3, k \geq 1$ are natural numbers. Finally, if $n \in \{1, 2\}$, then $f(1) = 3 < g(1) = 5 < 7 = f(2)$, and $g(2) = 31 = f(5)$. Therefore $S(3) \cap S(5) = \{31\}$.

Problem 1.54. If $k \geq 202$ and $n \geq 2k$, then prove that $C_n^k > n^{\pi(k)}$, where $\pi(k)$ denotes the number of prime numbers smaller than k.

Solution 1.54. 1. Let us show first that $C_{2k}^k > \frac{4^k}{k}$ if $k \geq 4$. If $k = 4$, then $C_{2k}^k = 70 \geq 64 = \frac{4^k}{k}$. We will use next a recurrence on k. We have

$$C_{2k+2}^{k+1} = \frac{(2k+2)(2k+1)}{(k+1)^2} C_{2k}^k > \frac{2(2k+1)}{k+1} \cdot \frac{4^k}{k} > \frac{4k \cdot 4^k}{k \cdot (k+1)} = \frac{4^{k+1}}{k+1}.$$

2. We will use the well-known estimate

$$\pi(k) < \frac{k}{\log k - \frac{3}{2}}, \quad \text{for } k > 5.$$

The inequality

$$\frac{4^k}{k} > (2k)^{k/(\log k - 3/2)}$$

is equivalent to $k \log 4 - \log k > k \frac{\log k + \log 2}{\log k - 3/2}$, which is true as soon as $k \geq 1414$. Moreover,

$$C_{2k}^k > \frac{4^k}{k} > 2^{kk/(\log k - 3/2)} > (2k)^{\pi(k)}, \quad \text{for } k \geq 1414.$$

By direct computer verification, one has $C_{2k}^k > (2k)^{\pi(h)}$ for $202 \leq h \leq 1413$.

3. Now if $s \geq 1$, we have

$$k > \pi(k) \geq 1 + \frac{\pi(k) - 1}{s} = \frac{\pi(k) + s - 1}{s}.$$

The product of these inequalities for $1 \leq s \leq r$ gives us $k^r > C_{\pi(k)+r-1}^r$. Then

$$C_{n+1}^k = \left(1 - \frac{k}{n+1}\right)^{-1} C_n^k > \left(1 - \frac{k}{n+1}\right)^{-1} n^{\pi(k)} = n^{\pi(k)} \cdot \sum_{r \geq 0}^{\infty} \frac{k^r}{(n+1)^r}$$

$$> n^{\pi(k)} \cdot \sum_{r \geq 0}^{\infty} \frac{C_{\pi(k)+r-1}^r}{(n+1)^r} = n^{\pi(k)} \left(1 - \frac{1}{n+1}\right)^{-\pi(k)} = (n+1)^{\pi(k)}.$$

Problem 1.55. Let $S_m(n)$ be the sum of the inverses of the integers smaller than m and relatively prime to n. If $m > n \geq 2$, then show that $S_m(n)$ is not an integer.

Solution 1.55. Let us write

$$S_m(n) = 1 + \frac{1}{a_1} + \cdots + \frac{1}{a_s},$$

where $1 < a_1 < a_2 < \cdots < a_s < m$ and $\gcd(a_i, n) = 1$. Observe first that $a = a_1$ is prime, since otherwise, any proper divisor of a would be smaller and still relatively prime to n. Let k be the largest integer such that $a^k \leq m$. Then there exists some $i \leq s$ such that $a^k = a_i$, because a^k is also a number relatively prime to n.

Moreover, if a^k divides a_j for some j, then $j = i$. In fact, let us assume that $a_j = ca^k$ for some integer $c > 1$. If $c < a$, then by hypothesis, $\gcd(c, n) > 1$, and hence a_j would not be relatively prime to n. If $c \geq a$, then $m > a_j = ca^k \geq a^{k+1}$, which would contradict the choice of k, supposed to be maximal with the property that $a^k \leq m$.

Let us suppose now that $S_m(n)$ is an integer, and set $L = \mathrm{lcm}(a_1, \ldots, a_s)$. One knows from above that a^k divides L, while a^{k+1} does not divide L.

We have also

$$LS_m(n) - \sum_{j=1, j \neq i}^{s} \frac{L}{a_j} = \frac{L}{a_i}.$$

If $S_m(n)$ were an integer, then the left-hand side would be a multiple of a. On the other hand, the term on the right-hand side is $\frac{L}{a_i} = \frac{L}{a^k}$, which cannot be a multiple of a, by the choice of k. This contradiction proves that $S_m(n)$ is not an integer.

Problem 1.56. Solve in integers the following equations:

1. $x^4 + y^4 = z^2$;
2. $y^3 + 4y = z^2$;
3. $x^4 = y^4 + z^2$.

Solution 1.56. 1. Suppose that there exists a nontrivial solution for which x or y is nonzero and hence a solution with $x, y, z > 0$. Consider the solution for which z is minimal among all possible solutions. We may suppose that $\gcd(x, y, z) = 1$. Then the triple (x^2, y^2, z) is a primitive Pythagorean triple and so $\gcd(x, y) = \gcd(y, z) = \gcd(x, z) = 1$. We can write $x^2 = m^2 - n^2$, $y^2 = 2mn$, $z = m^2 + n^2$, where m, n are integers. Since the solutions are primitive, it follows that x and z must be odd. In particular, $m \not\equiv n \pmod 2$, since otherwise, x would be even. We have two cases:

(i). If m is even and n is odd, then $x^2 = m^2 - n^2 \equiv -1 \pmod 4$, which is impossible, since any square is congruent to either 0 or 1 modulo 4.

(ii). If m is odd and n is even, then we have $y^2 = 2mn$ and $\gcd(m, n) = 1$. Thus $m = m_1^2$ and $n = 2n_1^2$. Moreover, $x^2 = m_1^4 - 4n_1^4$, and so

$$(m_1^2 - x)(m_1^2 + x) = 4n_1^4.$$

If d divides both $m_1^2 - x$ and $m_1^2 + x$, then d divides $2m$ and n and so $d = 2$ or $d = 1$. But $m_1^2 - x$ and $m_1^2 + x$ are congruent modulo 2, and they cannot be both odd, since their product is even. Thus $m_1^2 - x = 2a^4$ and $m_1^2 + x = 2b^4$. In particular, we find that

$$a^4 + b^4 = m_1^2,$$

where a, b are nonzero and $m_1 < m_1^2 = m < z$. Thus we have obtained a solution with z smaller than the initial one, which is a contradiction. Therefore there are no nontrivial solutions.

2. We have $(y^3 + 4y)^2 = y^6 + 8y^4 + 16y^2 = z^4$. Also, $(y^3 - 4y)^2 = y^6 + 16y^2 - 8y^4$. Adding these equations, we obtain $(2y)^4 = z^4 - (y^3 - 4y)^2$. It is known that the equation $u^4 - v^4 = w^2$ does not have nontrivial integer solutions (see the previous equation). Therefore either $2y = 0$ and $z^4 = (y^3 - 4y)^2 = 0$, which leads to $y = z = 0$, or else $z = 2y$, $y^3 - 4y = 0$, which implies $y = 2$, $z = 4$.

3. Consider a positive solution (x, y, z) with $\gcd(x, y) = 1$ and minimal x. Then (y^2, z, x^2) is a primitive Pythagorean triple, and hence we have two possible situations:

(i). Either $y^2 = m^2 - n^2, x^2 = m^2 + n^2, z = 2mn$, where $m > n$ and $\gcd(m, n) = 1$. Then

$$a^4 = b^4 + (xy)^2$$

and $a < x$, contradicting the fact that x was chosen minimal.

(ii). Or $z = m^2 - n^2, x^2 = m^2 + n^2, y^2 = 2mn$, where $m > n$ and $\gcd(m, n) = 1$. Then (m, n, x) is also a primitive Pythagorean triple and thus m is odd and n is even. Since $y^2 = 2mn$ and n is odd, we can write $m = 2a^2, n = b^2$, so that $y = 2ab$. Using the fact that (m, n, x) is a primitive Pythagorean triple, where m is even and n is odd, we derive that $m = 2rs, n = r^2 - s^2 = b^2$, with $\gcd(r, s) = 1$. Thus $a^2 = rs$, which yields $r = u^2, s = v^2$, and thus, expressing n as above, we obtain:

$$u^4 = v^4 + b^2$$

with $u < r < m < x$, contradicting the minimality of x. Thus there are no nontrivial solutions.

Comments 25 *In particular, the Fermat equation $x^4 + y^4 = z^4$ has no nontrivial solutions. The same method, called infinite descent, can be used successfully to prove that the equations $x^4 + dy^4 = z^2$ have no solutions, when d is prime and $d \equiv 7, 11 \pmod{16}$, or $\frac{d}{2}$ is prime and $\frac{d}{2} \equiv \pm 3 \pmod 8$, or $-d$ is prime and $-d \equiv \pm 3, 11 \pmod{16}$, or $\frac{d}{4}$ is prime and $\frac{d}{4} \equiv \pm 3, 11 \pmod 8$. For more information about Diophantine equations see*

- L.J. Mordell: *Diophantine Equations,* Academic Press, Pure and Applied Math. vol. 30, London, New York, 1969.

Problem 1.57. Show that the equation

$$x^3 + y^3 = z^3 + w^3$$

has infinitely many integer solutions. Prove that 1 can be written in infinitely many ways as a sum of three cubes.

Solution 1.57. 1. We make the change of variables $x + y = X, x - y = Y, z + w = Z$, $z - w = W$. The equation now reads

$$X\left(X^2 + 3Y^2\right) = Z\left(Z^2 + 3W^2\right).$$

We use the identity

$$\left(a^2 + 3b^2\right)\left(A^2 + 3B^2\right) = (aA + 3bB)^2 + 3(bA - aB)^2$$

and derive

$$X\left(X^2 + 3Y^2\right)^2 = Z\left((XZ + 3YW)^2 + 3(YZ - XW)^2\right).$$

Set

$$p = \frac{XZ + 3YW}{X^2 + 3Y^2}, \quad q = \frac{YZ - XW}{X^2 + 3Y^2},$$

and thus $\frac{X}{Z} = p^2 + 3q^2$. Choose $Z = 1$, for simplicity. We have then

$$pX + 3qY = 1, \quad pY - qX = W,$$

from which one can obtain the values of Y and W. Finally, we obtain

$$x = 1 - (p - 3q)\left(p^2 + 3q^2\right), \quad y = -1 + (p + 3q)\left(p^2 + 3q^2\right),$$
$$z = p + 3q - \left(p^2 + 3q^2\right)^2, \quad w = -p + 3q + \left(p^2 + 3q^2\right)^2.$$

These are all rational solutions of our equation. If $p, q \in \mathbb{Z}$, then we obtain infinitely many integer solutions. However, it is not clear that these formulas will provide all integer solutions of our equations.

2. Set $p = 3q, t = 2q$. We obtain

$$1 = \left(1 - 9t^3\right)^3 + \left(3t - 9t^4\right)^3 + \left(9t^4\right)^3.$$

Comments 26 *It is unknown whether* 3 *can be written in infinitely many ways as a sum of three cubes.*

Problem 1.58. Prove that the equation $y^2 = Dx^4 + 1$ has no integer solution, except $x = 0$, if $D \not\equiv 0, -1, 3, 8 \pmod{16}$ and there is no factorization $D = pq$, where $p > 1$ is odd, $\gcd(p, q) = 1$, and either $p \equiv \pm 1 \pmod{16}$, $p \equiv q \pm 1 \pmod{16}$, or $p \equiv 4q \pm 1 \pmod{16}$.

Solution 1.58. Let $x, y > 0$ be a nontrivial solution with minimal x. If x is odd, then $Dx^4 + 1 \not\equiv 0, 1, 4, 9 \pmod{16}$, and thus it is not a quadratic residue modulo 16. Thus x should be even and y odd. Further, we have

$$y + 1 = 2pa^4, \quad y - 1 = 8qb^4, \quad x = 2ab$$

or

$$y - 1 = 2pa^4, \quad y + 1 = 8qb^4, \quad x = 2ab,$$

where $D = pq$, $\gcd(p, q) = 1$, and a is odd.

If $p > 1$, then

$$pa^4 - 4qb^4 = \pm 1.$$

If b is even, then $p \equiv \pm 1 \pmod{16}$, contradiction. If b is odd, then $p \equiv 4q \pm 1 \pmod{16}$, again false.

Thus $p = 1$ and

$$a^4 - 4Db^4 = \pm 1.$$

The equation $a^4 - 4Db^4 = -1$ has no solution modulo 4; thus $a^4 - 1 = 4Db^4$. Then $D = rs$, $b = cd$, $\gcd(r, s) = 1$, and

$$a^2 + 1 = 2rc^4, \quad a^2 - 1 = 2sd^4.$$

Thus c, r are odd, and we have

$$rc^4 - sd^4 = 1.$$

If $r = 1$, then $c^4 - Dd^4 = 1$ and $x = 2acd$, so there exists a solution of our equation with $d \leq x/2$, contradicting the minimality of x.

If $r > 1$, then d cannot be even because $r \not\equiv \pm 1 \pmod{16}$. But if d is odd, then $r \equiv s + 1 \pmod{16}$, contradicting our assumptions.

Problem 1.59. Find all numbers that are simultaneously triangular, perfect squares, and pentagonal numbers.

Solution 1.59. We have the Diophantine equations

$$n = p^2 = \frac{1}{2}q(q + 1) = \frac{1}{2}r(3r - 1).$$

We derive that

$$(2q + 1)^2 = 1 + 8p^2 \text{and } (6r - 1)^2 = 1 + 24p^2.$$

In particular, we obtain

$$(6r - 1)^2 = (4p)^2 + (2q + 1)^2.$$

Since $\gcd(4p, 2q + 1) = 1$, the triple $(4p, 2q + 1, 6r - 1)$ is a primitive Pythagorean triple. This means that the integer solutions are determined as functions of two integer parameters $m, n \in \mathbb{Z}$ by means of the formulas

$$4p = 2mn, \quad 2q + 1 = m^2 + n^2, \quad 6r - 1 = m^2 + n^2.$$

Substituting in the previous equation, we obtain the $m^4 - 4m^2n^2 + n^4 = 1$ constraint on the parameters, which can also be written as $(m^2 - 2n^2)^2 = 3n^4 + 1$.

We now use a theorem that says that the Ljunggren equation

$$x^2 = Dy^4 + 1$$

has at most two positive integer solutions if $D > 0$ is not a perfect square.

In our case, the equation $x^2 = 3y^4 + 1$ has the obvious solutions $(7, 2)$ and $(2, 1)$ and thus these are all the solutions in natural numbers. Therefore there is only one solution $(m, n) = (2, 1)$.

Comments 27 *The proof above is based on a deep result concerning the equation* $x^2 = Dy^4 + 1$, *proved by Ljunggren in 1942. Ljunggren also gave an algorithm that computes the nontrivial solution when it exists. The previous problem is a particular case in which a short elementary proof (due to L. Mordell) is available.*

Ljunggren extended his methods to deal with the more-delicate equation $x^2 = Dy^4 - 1$. *He proved that this Diophantine equation has at most two positive solutions for* $D > 0$, *not a perfect square. In particular,* $x^2 = 2y^4 - 1$ *has only the solutions* $(1, 1)$ *and* $(13, 239)$, *which have been known for two centuries. Recently, Chen and Voutier proved that there is at most one solution in positive integers when* $D \geq 3$.

- W. Ljunggren: *Einige Eigenschaften der Einheiten reeller quadratischer und rein-biquadratischer Zahlkörper mit Anwendung auf die Lösung einer Klasse unbestimmter Gleichungen vierten Grades*, Skr. Norske Vid. Akad. Oslo 1936, 12, 1–73.
- W. Ljunggren: *Über die Gleichung* $x^4 - Dy^2 = 1$, Arch. Math. Naturvid. 45 (1942) 5, 61–70.
- W. Ljunggren: *Zur Theorie der Gleichung* $x^2 + 1 = Dy^4$, Avh. Norske Vid. Akad. Oslo. I. 1942.
- J.H. Chen and P. Voutier: *Complete solution of the Diophantine equation* $X^2 + 1 = dY^4$ *and a related family of quartic Thue equations*, J. Number Theory 62 (1997), 1, 71–99.

Problem 1.60. Find all inscribable integer-sided quadrilaterals whose areas equal their perimeters.

Solution 1.60. We have $16S^2 = \prod_{\text{cyclic}}(a + b + c - d)$, and so the problem is equivalent to the Diophantine equation

$$16(a + b + c + d)^2 = \prod_{\text{cyclic}} (a + b + c - d).$$

Assume that $0 < d \leq c \leq b < a \leq b + c + d$, for the sake of simplicity. If we write $x = -a + b + c + d$, $y = a - b + c + d$, $z = a + b - c + d$, $t = a + b + c - d$, then the above equation becomes

$$(x + y + z + t)^2 = xyzt.$$

If this equation has a real solution, then the quadratic equation $(X + y + z + x)^2 - Xxyz = 0$ has two real roots. The product of these real roots is $(x + y + z)^2$. Since the solution t satisfies $0 < t < x + y + z$, it follows that t is the smallest of the two roots, and thus it is determined by the formula

$$2t = xyz - 2(x + y + z) - \sqrt{xyz(xyz - 4x - 4y - 4z)}$$

with the obvious constraint $xyz > 4(x + y + z)$.

Therefore, we need that $\lambda = \frac{x+y+z}{xyz} < \frac{1}{4}$. Now, $3t \geq x + y + z$, since t is the longest side, and thus $\frac{8}{3}\lambda + \sqrt{1 - 4\lambda} \leq 1$, implying that $\lambda \geq \frac{3}{16}$.

We have obtained the inequalities $\frac{1}{4} > \lambda \geq \frac{3}{16}$, which immediately imply that $5 \leq xy \leq 16$.

Since $\lambda < \frac{1}{4}$, we get $z > 4(x + y)/(xy - 4)$.

On the other hand, we assumed that $z \leq t$, which is equivalent to

$$(xy - 4)z^2 - 4(x + y)z - (x + y)^2 \leq 0,$$

and therefore $z \leq (x + y)/(\sqrt{xy} - 2)$. Consequently,

$$\frac{4(x + y)}{xy - 4} \leq z \leq \frac{x + y}{\sqrt{xy} - 2}.$$

One can list all integer values of x, y, z, t satisfying the inequalities above and that are solutions to the Diophantine equation. This leads to

$$(x, y, z, t) \in \{(1, 9, 10, 10), (2, 5, 5, 8), (3, 3, 6, 6), (4, 4, 4, 4)\}$$

and their permutations.

Problem 1.61. Every rational number a can be written as a sum of the cubes of three rational numbers. Moreover, if $a > 0$, then the three cubes can be chosen to be positive.

Solution 1.61. Assume that we know how to write $a = x^3 + y^3 + z^3$ and let us introduce $u = x+y+z$ and $v = x+y$. Then $a = (x+y+z)^3 - 3(x+y)(y+z)(z+x)$, and we derive easily that $a = u^3 - 3v(u - x)(u - v + x)$.

We will require furthermore that u and v satisfy the additional constraint $u^3 = 3v(u^2 - x^2)$. Then $a = 3v^2(u - x)$.

We introduce the new variable $w = \frac{x}{u} \in \mathbb{Q}$, so that $u = 3v(1 - w^2)$. Therefore $a = 3v^2(u - x) = 3v^2u(1 - w) = 9v^3(1 - w)^2(1 + w) = 27v^3(1 - w)^3 \cdot \frac{1}{3} \cdot \frac{1+w}{1-w}$.

Set $m = 3v(1 - w)$ and introduce it above in order to obtain $a = \frac{m^3}{3}\frac{1+w}{1-w}$.

Now we reverse the line of thought by letting $m \in \mathbb{Q}$ be an arbitrary parameter. The previous formulas yield $w = \frac{3a - m^3}{3a + m^3}$ and also the following solutions:

$$v = \frac{3a + m^3}{6m^2}, \quad u = \frac{6am}{3a + m^3}.$$

This yields the rational solutions $x = wu, y = v - wu, z = u - v$.

Let us prove the positivity part. We have that $x = uw \geq 0$ is equivalent to $0 < m < \sqrt[3]{3a}$. But let us assume for a moment that $m = \sqrt[3]{3a}$. Then $y = \frac{\sqrt[3]{3a}}{3}, z = \frac{2\sqrt[3]{3a}}{3}, x = 0$. Regarding the solutions y, z as functions $y = y(m), z = z(m)$ of the parameter m, we immediately see that these are continuous functions, which are strictly positive at $m = \sqrt[3]{3a}$. In particular, there exists a small neighborhood W of $\sqrt[3]{3a}$ such that $y(m), z(m) > 0$. This means that for $m \in \mathbb{Q}_+ \cap (0, \sqrt[3]{3a}) \cap W$, we have $x, y, z \in \mathbb{Q}_+$, as claimed.

Problem 1.62. 1. Every prime number of the form $4m + 1$ can be written as the sum of two perfect squares.

2. Every natural number is the sum of four perfect squares.

Solution 1.62. 1. Let p be such a prime number. The congruence $z^2 + 1 \equiv 0 \pmod{p}$ has at least one solution, because -1 is a quadratic residue when p is of the form $4m + 1$. This follows from the more general Euler criterion computing the Legendre symbol $\left(\frac{a}{p}\right)$ as follows:

$$\left(\frac{a}{p}\right) = a^{\frac{p-1}{2}} \pmod{p}$$

for arbitrary prime p. Recall that the Legendre symbol $\left(\frac{a}{b}\right)$ equals 1 if a is a quadratic residue modulo b (i.e., there exists some integer n such that $n^2 \equiv a \pmod{b}$), and -1 otherwise.

Let then $m \in \mathbb{Z}_+$, $m \neq 0$, and $z \in \mathbb{Z}$ be such that

$$mp = z^2 + 1.$$

We can suppose that $-\frac{p}{2} < z < \frac{p}{2}$, by using a translation multiple of p. Then $m = \frac{1}{p}(z^2 + 1) < \frac{1}{p}\left(\frac{p^2}{4} + 1\right) < p$. Set $x = z$, $y = 1$, so that $mp = x^2 + y^2$.

We shall prove that if $m > 1$, then there exists $m' < m$ with the same properties. This process stops when $m = 1$, where we find the wanted sum of two squares.

Let us consider u, v such that $-\frac{m}{2} \leq u, v \leq \frac{m}{2}$, satisfying

$$u \equiv x \pmod{m}, \quad v \equiv y \pmod{m}.$$

Thus

$$u^2 + v^2 = x^2 + y^2 \equiv 0 \pmod{m},$$

and so there exists some $r \in \mathbb{Z}_+$ for which

$$u^2 + v^2 = mr.$$

Observe that $r \neq 0$, since otherwise x, y would be multiples of m, and so m would divide p. Further, $r = \frac{1}{m}\left(u^2 + v^2\right) < \frac{1}{m}\left(\frac{m^2}{4} + \frac{m^2}{4}\right) < m$. We will use the identity

$$\left(a^2 + b^2\right)\left(A^2 + B^2\right) = (aA + bB)^2 + (aB - bA)^2$$

and multiply the two equalities above, obtaining

$$m^2 rp = \left(x^2 + y^2\right)\left(u^2 + v^2\right) = (xu + yv)^2 + (xv - yu)^2.$$

Moreover,

$$xu + yv \equiv x^2 + y^2 \equiv 0 \pmod{m} \quad \text{and} \quad xv - yu \equiv xy - yx \equiv 0 \pmod{m},$$

and thus, dividing the equation by m^2, we obtain

$$rp = a^2 + b^2,$$

where $r < m$. Thus, $m' = r$ satisfies all required properties. This proves the first claim.

2. The identity

$$(a^2 + b^2 + c^2 + d^2)(A^2 + B^2 + C^2 + D^2) = (aA + bB + cC + dD)^2$$

$$+(aB - bA - cD + dC)^2 + (aC + bD - cA - dB)^2 + (aD - bC + cB - dA)^2$$

reduces the problem to the case of prime numbers. Moreover, we know that primes of the form $4m + 1$ can be written as a sum of two squares, and $2 = 1^2 + 1^2$. Thus it suffices to consider the primes of the form $4m + 3$. The method of proof is the same as above.

We show first that the congruence $x^2 + y^2 + 1 \equiv 0 \pmod{p}$ has solutions. By the Euler criterion, -1 is a quadratic nonresidue modulo p and thus $-y^2$ is a quadratic nonresidue modulo p. Moreover, each quadratic nonresidue is of the form $-y^2 \pmod{p}$. Then we have to find a quadratic nonresidue n and a quadratic residue r such that $r + 1 = n \pmod{p}$. Consider in the sequence $\{1, 2, 3, \ldots, p - 1, p\}$ the smallest quadratic nonresidue n. Then $n > 1$ and $n - 1$ must be a quadratic residue by the minimality of n, and this yields our pair (r, n).

We write now

$$mp = x^2 + y^2 + 1^2 + 0^2,$$

where $m \in \mathbb{Z}_+$ is nonzero. We choose x, y such that $-\frac{p}{2} < x, y < \frac{p}{2}$, using a translation multiple of p. Then $m = \frac{1}{p}(x^2 + y^2 + 1) < \frac{1}{p}\left(\frac{p^2}{4} + \frac{p^2}{4} + 1\right) < p$. Consider $a = x, b = y, c = 1, d = 0$. If $m > 1$, we will find another $m' < m$ such that $m'p$ is still a sum of four squares.

Choose A, B, C, D such that $-\frac{m}{2} < A, B, C, D \leq \frac{m}{2}$ and

$$a \equiv A \pmod{m}, \quad b \equiv B \pmod{m}, \quad c \equiv C \pmod{m}, \quad d \equiv D \pmod{m}.$$

Then $A^2 + B^2 + C^2 + D^2 \equiv 0 \pmod{m}$ and thus there exists $r \in \mathbb{Z}_+$ such that

$$mr = A^2 + B^2 + C^2 + D^2.$$

Furthermore, $r \neq 0$, since otherwise, $A = B = C = D = 0$, and thus a, b, c, d would be multiples of m, and so m would divide p. Moreover, $r = \frac{1}{m}(A^2 + B^2 + C^2 + D^2) \leq m$, with equality if and only if $A = B = C = D = \frac{m}{2}$, m even. The equality would imply that $a^2 = \frac{m^2}{4} \pmod{p}$ and thus $pm \equiv 0 \pmod{m^2}$. Thus m divides p, which is false. Thus $r < m$. Further, multiplying the two identities, we obtain

$$m^2 rp = \left(a^2 + b^2 + c^2 + d^2\right)\left(A^2 + B^2 + C^2 + D^2\right) = X^2 + Y^2 + Z^2 + W^2.$$

The terms X, Y, Z, W are multiples of m, since, for example,

$$X = aA + bB + cC + dD \equiv a^2 + b^2 + c^2 + d^2 \equiv 0 \ (\text{mod } m),$$

and similarly for Y, Z, W. Therefore, one can divide by m^2 and find that

$$rp = X'^2 + Y'^2 + Z'^2 + W'^2$$

with $r < m$. Then take $m' = r$. This proves the claim.

Comments 28 *The proof given above was obtained by Lagrange, and all elementary proofs known until now are variations of the present one. The numbers of the form $4^m (8k + 7)$ cannot be represented as sums of three squares. Thus 4 is the minimal number with this property. However, all numbers that are not of the form $4^m (8k + 7)$ are sums of three perfect squares.*

Problem 1.63. Every natural number can be written as a sum of at most 53 integers to the fourth power.

Solution 1.63. Consider the identity

$$6\left(x_1^2 + x_2^2 + x_3^2 + x_4^2\right)^2 = \sum_{1 \le i \le j \le 4} (x_i - x_j)^4 + \sum_{1 \le i < j \le 4} (x_i + x_j)^4.$$

We write $n = 6x + y$, where $0 \le y \le 5$. According to Lagrange's theorem (the previous problem), any natural number is the sum of four perfect squares. Therefore

$$x = \alpha_1^2 + \alpha_2^2 + \alpha_3^2 + \alpha_4^2, \quad \alpha_i = \alpha_{i_1}^2 + \alpha_{i_2}^2 + \alpha_{i_3}^2 + \alpha_{i_4}^2.$$

Then

$$6x = \sum_{i=1}^{4} 6\left(\sum_{j=1}^{4} \alpha_{i_j}^2\right)^2$$

$$= \sum_{i=1}^{4} \sum_{1 \le j < k \le 4} (\alpha_{i_j} - \alpha_{i_k})^4 + \sum_{i=1}^{4} \sum_{1 \le j < k \le 4} (\alpha_{i_j} + \alpha_{i_j} + \alpha_{i_k})^4 = \sum_{i=1}^{48} w_i^4.$$

Now, any $y \in \{0, \ldots, 5\}$ can be written as a sum of at most 5 fourth powers, and thus n can be written as a sum of at most 53 fourth powers.

Comments 29 *The problem of determining the minimum $G(k)$ such that any number is the sum of $G(k)$ perfect (positive) kth powers (in our case $k = 4$) is called the Waring problem. For instance, Lagrange proved in 1770 that $G(2) = 4$. Earlier the same year, Edward Waring had conjectured that something similar could be proved for cubes, fourth powers, and so on. He stated, without proof, that it would take the sum of at most nine cubes or 19 fourth powers to express any positive integer. Waring arrived at his conjectures about cubes and fourth powers by collecting data and looking for patterns. The sequence of minimum number of cubes needed to write $n \le 30$ as a sum of cubes is* 1, 2, 3, 4, 5, 6, 7, 1, 2, 3, 4, 5, 6, 7, 8, 2, 3, 4, 5, 6, 7, 8, 9, 2, 3, 4, 1, 2, 3, 4.

Fourth powers show similar behavior: 15 *can be written as the sum of* 15 *fourth powers,* 31 *requires* 16 *fourth powers,* 47 *requires* 17, 63 *requires* 18, *and* 79 *requires* 19. *Among the first* 12000 *numbers, the only number other than* 23 *that requires nine cubes is* 239. *Fifteen numbers require a minimum of eight cubes:* 15, 22, 50, 114, 167, 175, 186, 212, 213, 238, 303, 364, 420, 428, *and* 454. *The list of numbers requiring seven cubes is much longer, but it includes no numbers greater than* 8042.

In 1909, David Hilbert proved that there is some universal constant $G(k)$ such that every natural number is the sum of at most $G(k)$ kth powers. However, his proof did not provide an effective $G(k)$. In 1912, A.J. Kempner completed a 1909 partial proof by A. Wieferich to establish, once and for all, that every integer can be expressed as a sum of nine cubes. In 1940, S.S. Pillai showed that every integer can be expressed as a sum of 73 sixth powers.

The easy estimate obtained above, that $G(4) \leq 53$, is far from being sharp. The assertion that 37 fifth powers are sufficient was proved by Chen Jingrun in 1964. It wasn't until 1986 that R. Balasubramanian, J.-M. Deshouillers, and F. Dress proved that $G(4) = 19$, by showing that it suffices to check the claim for those numbers less than 10^{367}.

A related question concerns the number of terms required to express every sufficiently large integer as a sum of kth powers. For example, even though every integer can be expressed as a sum of at most nine cubes, every integer larger than a certain value (probably 8 042) can be written as the sum of at most seven cubes. One conjectures that every sufficiently large integer can be expressed as the sum of no more than four cubes. The largest number now known not to be a sum of four cubes is 7 373 170 279 850. Davenport proved that any sufficiently large integer is represented as the sum of 16 fourth powers, and Kempner that no integer of the form $16^h 31$ is representable as a sum of fewer than 16 integral fourth powers.

In 1995, Irving Kaplansky and Noam D. Elkies proved that any integer can be expressed as the sum of two squares and a cube, when positive and negative integers are allowed. Numerical experiments show evidence for the conjecture that all but finitely many numbers can be expressed as the sum of a square and two cubes, using only positive integers. See also:

- R. Balasubramanian, J.-M. Deshouillers, and F. Dress: *Problème de Waring pour les bicarrés. I. Schéma de la solution.* C. R. Acad. Sci. Paris Sér. I Math. 303 (1986), 4, 85–88.
- H. Davenport: *On Waring's problem for fourth powers,* Ann. of Math. (2) 40 (1939), 731–747.
- W.C. Jagy and I. Kaplansky: *Sums of squares, cubes, and higher powers,* Experimental Mathematics 4 (1995), 169–173.
- H. Rademacher and O. Toeplitz: *The Enjoyment of Mathematics: Selections from Mathematics for the Amateur,* New York, Dover, 1990.

Problem 1.64. Let $G(k)$ denote the minimal integer n such that any positive integer can be written as the sum of n positive perfect kth powers. Prove that $G(k) \geq 2^k + \left[\frac{3^k}{2^k}\right] - 2$.

Solution 1.64. Let us write $3^k = 2^k q + r$, where $0 < r < 2^k$ and $q = \left[\frac{3^k}{2^k}\right]$, and consider the number $A = 2^k q - 1$. We can write $A = (q-1)2^k + (2^k - 1)1^k$, thus as a sum of $2^k + \left[\frac{3^k}{2^k}\right] - 2$ perfect kth powers. Let us show that we cannot do better. First, $A < 3^k$, and thus we cannot use 3^k among our kth powers. On the other hand, $A < 2^k q$, and thus at most $q - 1$ terms among our kth powers could be equal to 2^k. The remaining terms should be 1^k, and thus there are at least $q - 1 + 2^k - 1$, as claimed.

Comments 30 *This is the general case of the Waring problem, discussed above. It is conjectured that $G(k) = 2^k + \left[\frac{3^k}{2^k}\right] - 2$, for any $k \geq 2$. Moreover, the conjecture was proved by L.E. Dickson and (independently) by S.S. Pillai for those k satisfying*

$$r = 3^k - 2^k q \leq 2^k - q - 2,$$

where q and r are as above. This inequality was verified for all $k \leq 471\,600\,000$. Moreover, it is known that the inequality holds for all sufficiently large k, but we do not have an effective lower bound.

Problem 1.65. Let $R(n, k)$ be the remainder when n is divided by k and

$$S(n, k) = \sum_{i=1}^{k} R(n, i).$$

1. Prove that $\lim_{n \to \infty} \frac{S(n, n)}{n^2} = 1 - \frac{\pi^2}{12}$.
2. Consider a sequence of natural numbers (a_k) growing to infinity and such that $\lim_{k \to \infty} \frac{a_k \log k}{k} = 0$. Prove that $\lim_{k \to \infty} \frac{S(ka_k, k)}{k^2} = \frac{1}{4}$.

Solution 1.65. 1. We calculate that:

$$S(n+1, n+1) = S(n, n) - \sigma(n+1) + 2n + 1,$$

where $\sigma(m)$ denotes the sum of divisors of m. Therefore, we find that $S(n, n) = n^2 - \sum_{j \leq n} \sigma(j)$. Furthermore, use the estimate

$$\sum_{j \leq n} \sigma(j) = \frac{1}{2}\zeta(2)n^2 + \mathcal{O}(n \log n)$$

and get the claim.

2. We have the identities

$$S(n, k) = nk - \sum_{i=1}^{k} \left[\frac{n}{i}\right] i = nk - \left[\frac{n}{k}\right] \sum_{i=1}^{k} i - \sum_{\frac{n}{k} < k \le n} \left(\sum_{i=1}^{\left[\frac{n}{k}\right]} i\right)$$

in order to obtain the estimate

$$S(n, k) = nk - \left[\frac{n}{k}\right] \frac{k(k+1)}{2} - \frac{n^2}{2} \sum_{i=\left[\frac{n}{k}\right]+1}^{\infty} \frac{1}{i^2} + \mathcal{O}(n \log k).$$

This automatically yields the first claim. Further,

$$S(ka_k, k) = \frac{1}{2}\left(k^2 a_k - ka_k - k^2 a_k^2 \sum_{i=a_k+1}^{\infty} \frac{1}{i^2}\right) + \mathcal{O}(ka_k \log k),$$

and we can estimate

$$\left(\sum_{i=N+1}^{\infty} \frac{1}{i^2}\right) - \frac{1}{N + \frac{1}{2}}$$

$$= \sum_{i=N+1}^{\infty} \frac{1}{i^2} - \sum_{i=N+1}^{\infty} \left(\frac{1}{N - \frac{1}{2}} - \frac{1}{N + \frac{1}{2}}\right)$$

$$= -\frac{1}{4} \sum_{i=N+1}^{\infty} \frac{1}{i^4 - \frac{i^2}{4}} = \mathcal{O}\left(\frac{1}{N^3}\right),$$

and hence

$$\frac{S(ka_k, k)}{k^2} = \frac{1}{4} \frac{a_k}{a_k + \frac{1}{2}} - \frac{1}{2k} + \mathcal{O}(a_k^{-1}) + \mathcal{O}\left(\frac{a_k \log k}{k}\right),$$

which proves the claim.

Comments 31 *L. Funar conjectured in 1985 that* $\lim_{n \to \infty} \frac{S(2^n, n)}{n^2} = 1 - \frac{\pi^2}{12}$. *This question seems to be still open.*

- L. Funar: *Problem 6476,* Amer. Math. Monthly, 81(1984), 588.

Problem 1.66. Let $a_1 < a_2 < a_3 < \cdots < a_n < \cdots$ be a sequence of positive integers such that the series

$$\sum_{i=1}^{\infty} \frac{1}{a_i}$$

converges. Prove that, for any i, there exist infinitely many sets of a_i consecutive integers that are not divisible by a_j for all $j > i$.

Solution 1.66. Fix an arbitrary i. The set $S_n = \{n - a_i, n - a_i + 1, \ldots, n - 1\}$ is called admissible for a_j (where $j > i$) if no element of S_n is divisible by a_j. It is obvious that S_n is admissible for a_j if a_j divides n. Moreover, S_n is not admissible for a_j if there exists some e, with $1 \leq e \leq a_i$, such that $n \equiv e \pmod{a_j}$.

Consider $\varepsilon > 0$ sufficiently small and k sufficiently large such that

$$\sum_{j>k} a_j^{-1} < \varepsilon.$$

Let $\beta = a_{i+1} \cdots a_k$ and S_n with the property that β divides n. Observe first that the sets S_n are admissible for those a_j with $i \leq j \leq k$. Now assume that $j > k$. Let $C_j(x)$ denote the set of those $c \leq x$ for which $S_{c\beta}$ is not admissible for a_j. Recall that $c \in C_j(x)$ iff there exists $e \in \{1, 2, \ldots, a_i\}$ such that $c\beta \equiv e \pmod{a_j}$.

If $\gcd(\beta, a_j) = d$, then d should divide e, and we can write $\beta = d\beta', e = de', a_j = da_j'$. Moreover, the condition above is equivalent to $c\beta' \equiv e' \pmod{a_j'}$.

If $a_j > \beta x$, then obviously $C_j(x) = \emptyset$. Further, for a fixed value of e', there are at most $\frac{x}{a_j} + 1$ solutions c for the previous congruence in the given range. Since e' can take at most $\frac{a_i}{d}$ values, we derive that the cardinality of $C_j(x)$ is bounded by $a_i \left(\frac{x}{a_j} + 1 \right)$.

Therefore, among the multiples of β satisfying $n < \beta x$, there are at most

$$\sum_{j>k} \frac{a_i x}{a_j} + a_i \operatorname{card}\{j | a_j < \beta x\} = \varepsilon a_i x + o(x)$$

integers n such that S_n is not admissible. In fact, the estimate $\operatorname{card}\{j | a_j < \beta x\} = o(x)$ is a consequence of the convergence of the series. This asymptotic estimate implies that for large x there exist infinitely many values of n that do not belong to any C_j, so that S_n is admissible with respect to all a_j.

Problem 1.67. Denote by C_n the claim that there exists a set of n consecutive integers such that no two of them are relatively prime. Prove that C_n is true for every n, such that $17 \leq n \leq 10000$.

Solution 1.67. We suppose that both p and $2p + 1$ are primes and consider those n for which $3p + 2 \leq n \leq p^2$. Set q for the product $\prod_{p_i \leq n, p_i \neq p, 2p+1} p_i$ of those prime numbers smaller than n and distinct from p and $2p + 1$. From the Chinese remainder theorem, there exists a natural number x satisfying the congruences

$$x \equiv 0 \pmod{q}, \quad x \equiv -3p - 1 \pmod{p(2p + 1)}.$$

Now let $S_n = \{x, x + 1, \ldots, x + n - 1\}$. Then S_n contains $x + 1, x + 2p + 1, \ldots, x + kp + 1, \ldots$, which have p as a common factor. Also, $x + p$ has the factor $2p + 1$ in common with $x + 3p + 1 \in S_n$. All other numbers from S_n have a common divisor with q, therefore with x. Therefore the set S_n satisfies the claim C_n, and we have to look for specific values of p.

If $p = 5$, then S_n holds for $17 \leq n \leq 25$. We choose further $p = 11, 29, 251, 1013,$ $49919, 5008193$, and this yields the claim for all n such that $35 \leq n \leq (5008193)^2$.

The case $26 \leq n \leq 34$ must be treated separately. There one considers x satisfying

$$x \equiv 0 \pmod{2 \cdot 5 \cdot 11 \cdot 17}, \quad x \equiv -1 \pmod 3, \quad x \equiv -2 \pmod{7 \cdot 19},$$
$$x \equiv -3 \pmod{13}, \qquad\qquad x \equiv -4 \pmod{23}.$$

Comments 32 *The claim C_n is false for all $n \leq 16$, and true for all $n \geq 17$, as proved in:*

1. S.S. Pillai: *On m consecutive integers,* Proc. Indian Acad. Sci., Sect. A. 11 (1940), 6–12, 13 (1941), 530–533.
2. A. Brauer: *On a property of k consecutive integers,* Bull. Amer. Math. Soc. 47 (1941), 328–331.

The proof, which is elementary, is based on the estimate

$$\pi(2n) - \pi(n) \geq 2 \left[\frac{\log n}{\log 2} \right] + 2, \text{ for } n \geq 75.$$

Problem 1.68. Prove that a natural number n has more divisors that can be written in the form $3k + 1$, for $k \in \mathbb{Z}$, than divisors of the form $3m - 1$, for $m \in \mathbb{Z}$.

Solution 1.68. Let $f(n)$ (and respectively $g(n)$) be the number of divisors of the form $3k + 1$ (and respectively $3m - 1$), for $k, m \in \mathbb{Z}$. Let $n = uv$, where $\gcd(u, v) = 1$. By making use of the obvious congruences

$$(-1)(-1) \equiv 1 \pmod 3, \quad (-1)1 \equiv -1 \pmod 3,$$

we obtain the formulas

$$f(n) = f(u)f(v) + g(u)g(v), \quad g(n) = g(u)f(v) + f(u)g(v).$$

Set further $h(n) = f(n) - g(n)$. The identities above prove that

$$h(n) = h(u)h(v).$$

In particular, if $n = p_1^{\alpha_1} \cdots p_r^{\alpha_r}$ is the factorization into prime factors of n, then

$$h(n) = \prod_{j=1}^r h\left(p_j^{\alpha_j}\right).$$

Now we have
$$h(3^\alpha) = 1.$$

If the prime number $p \equiv -1 \pmod 3$, then the divisors of p^α of the form $3k + 1$ are $1, p^2, p^4, \ldots$ and the ones of the form $3k - 1$ are $p, p^3, \ldots$. Therefore

$$h(p^\alpha) = \begin{cases} 1, & \text{if } \alpha \text{ is even} \\ 0, & \text{otherwise.} \end{cases}$$

If the prime number p satisfies $p \equiv 1 \pmod 3$, then all divisors of p^α are of the form $3k + 1$, and therefore $h(p^\alpha) = 1 + \alpha$. Therefore if $n = p_1^{\alpha_1} \cdots p_r^{\alpha_r}$,

$$h(n) = \begin{cases} \prod_{\nu;\, p_\nu \equiv 1 \pmod 3}(1 + \alpha_\nu), & \text{if all } \alpha_\nu \text{ for which } p_\nu \equiv 1 \pmod 3 \text{ are even,} \\ 0, & \text{otherwise.} \end{cases}$$

Comments 33 *It can be proved that $6h(n)$ is the number of integer solutions of the equation*

$$x^2 - xy + y^2 = n.$$

In the same way, we can prove that the difference between the number $\tau_{4k+1}(n)$ of divisors of the form $4k + 1$ and the number $\tau_{4k-1}(n)$ of divisors of the form $4k - 1$ is also positive. Moreover, $4(\tau_{4k+1}(n) - \tau_{4k-1}(n))$ is the number of solutions of the Diophantine equation

$$x^2 + y^2 = n.$$

Comments 34 *The positivity result holds more generally when one compares the numbers of divisors of the form $4k \pm 1$, $6k \pm 1$, $8k \pm 1$, $12k \pm 1$, $24k \pm 1$. However, the analogous result do not hold for any other congruences.*

Comments 35 *The number of solutions in integers of quadratic equations appears in many arithmetic questions. The oldest and perhaps one of the most famous unsolved problems is the congruent numbers problem. Specifically, one asks to determine all integers that are the areas of right triangles with rational sides, which are called congruent numbers. For instance, $n = 6$ is the area of the right triangle of sides 3, 4, and 5. It is less obvious that 157 is a congruent number, where the simplest solution for the sides of the associated right triangle reads*

$$\frac{6803298487826435051217540}{411340519227716149383203}, \quad \frac{411340519227716149383203}{21666555693714761309610},$$

$$\frac{224403517704336969924557513090674863160948472041}{8912332268928859588025535178967163570016480830}.$$

It is known that congruent numbers are those d for which the equation

$$dy^2 = x^3 - x$$

has infinitely many rational solutions.

Deep results of Tunnell imply the following conjecture, which would be a complete characterization of congruent numbers. Let d be square-free. Denote by $N(d)$ (and respectively $M(d)$) the number of integer solutions of

$$2x^2 + y^2 + 8z^2 = d, \text{ and respectively } 2x^2 + y^2 + 32z^2 = d, \text{ for odd } d,$$

$$4x^2 + y^2 + 8z^2 = \frac{d}{2}, \text{ and respectively } 4x^2 + y^2 + 32z^2 = \frac{d}{2}, \text{ for even } d.$$

Then d is a congruent number if and only if

$$N(d) = 2M(d).$$

This is a particular case of the Birch–Swinnerton-Dyer conjecture, considered one of the most important unsolved problems in arithmetic, and included as one of the seven Millennium Problems (for which the Clay Mathematics Institute will award one million dollars for a solution). The interested reader might consult

- N. Koblitz: *Introduction to Elliptic Curves and Modular Forms,* second edition, Graduate Texts in Mathematics, 97, Springer-Verlag, New York, 1993.

Problem 1.69. A number N is called deficient if $\sigma(N) < 2N$ and abundant if $\sigma(N) > 2N$.

1. Let k be fixed. Are there any sequences of k consecutive abundant numbers?
2. Show that there are infinitely many 5-tuples of consecutive deficient numbers.

Solution 1.69. 1. Let p_n denote the nth prime number. It is known that the infinite product

$$\prod_{n=1}^{\infty} \left(1 + \frac{1}{p_n}\right)$$

is infinite. Thus, for any m, there exists an integer $N(m)$ such that

$$\prod_{n=m}^{N(m)} \left(1 + \frac{1}{p_n}\right) > 2.$$

In particular, this shows that the number $t = \prod_{n=m}^{N(m)} p_n$ is abundant, because $\sigma(t) = t \prod_{n=m}^{N(m)} \left(1 + \frac{1}{p_n}\right)$. Consider now the sequence defined by

$$m_1 = 1, \ m_{j+1} = N(m_j) + 1, \text{ for } j \geq 1, \text{ and } t_j = \prod_{n=m_j}^{N(m_j)} p_n.$$

Then the numbers t_j are pairwise relatively prime, and therefore for every k there exists some integer L (depending on k) with the property that $L + j \equiv 0 \pmod{t_j}$, for all $j \in \{1, \ldots, k\}$. Observe that any multiple of an abundant number is abundant itself. This shows that each of the numbers $L + 1, \ldots, L + k$ is abundant.

2. It is well known that

$$\lim_{n \to \infty} \left(1 + \frac{2}{n \log n}\right)^n = 1$$

and $\lim_{n \to \infty} \frac{p_n}{n \log n} = 1$. Take n large enough so that

$$\left(1 + \frac{2}{n \log n}\right)^n < 1.01, \text{ and } p_n > \frac{1}{2}n \log n > 60.$$

Let us next consider

$$Q_n = \prod_{i=1}^{n-1} p_j, \quad \text{and} \quad M_n = 60p_n + 1.$$

We set $x_i = \frac{M_n + i - 1}{i}$ for $i \in \{1, 2, 3, 4, 5\}$. Because $n > 3$, it is easy to see that every x_i is relatively prime to Q_n. Therefore each x_i can be factored into prime factors

$$x_i = \prod_{j=1}^{t_i} p_{n_{ji}},$$

where each $n_{ji} \geq n$ (because otherwise, it would have common factors with Q_n) and the number of factors t_i is less than or equal to n (otherwise, x_i would be greater than M_n).

Since $1 + \frac{1}{p_{n_{ji}}} \leq 1 + \frac{1}{p_n}$ and for prime p we have $\sigma(p^s)/p^s \leq (\sigma(p)/p)^s$, we conclude that

$$\frac{\sigma(x_i)}{x_i} \leq \left(1 + \frac{1}{p_n}\right)^n < \left(1 + \frac{2}{n \log n}\right)^n < 1.01$$

and therefore

$$\frac{\sigma(M_n + i - 1)}{M_{n+i}} = \frac{\sigma(i)}{i} \cdot \frac{\sigma(x_i)}{x_i} < 1.75 \cdot 1.01 < 2.$$

Thus $M_n, M_n + 1, M_n + 2, M_n + 3, M_n + 4$ are deficient numbers.

Problem 1.70. Does there exist a nonconstant polynomial $an^2 + bn + c$ with integer coefficients such that for any natural number m, all its prime factors p_i are congruent to to 3 modulo 4? Prove that, for any nonconstant polynomial f with integer coefficients and any $m \in \mathbb{Z}$ there exist a prime number p and a natural number n such that p divides $f(n)$ and $p \equiv 1 \pmod{m}$.

Solution 1.70. 1. We write $4af(n) = (2an+b)^2 + 4ac - b^2$. If $a = 0$ or $4ac - b^2 = 0$, it is obvious that there are infinitely many prime numbers $4m+1$ that divide $f(n)$. Let us then assume that $a(4ac - b^2) \neq 0$. Set $p_1, \ldots, p_m$ for the odd prime divisors of $a(4ac - b^2)$. The Chinese remainder theorem and Dirichlet's theorem on arithmetic progressions imply the existence of a prime p satisfying $p \equiv 1 \pmod{p_i}$, for all i and $p \equiv 1 \pmod 8$. The Legendre symbol satisfies the identities $\left(\frac{p_i}{p}\right) = \left(\frac{p}{p_i}\right) = \left(\frac{2}{p}\right) = \left(\frac{-1}{p}\right) = 1$, which imply that there exists an integer m such that $m^2 \equiv b^2 - 4ac \pmod p$. Since $\gcd(2a, p) = 1$, the congruence $2an + b \equiv m \pmod p$ has some solution n and then p divides $f(n)$. The answer is therefore negative.

2. Assume that f is irreducible. Let $\alpha_1, \ldots, \alpha_k$ be the roots of f and let ω be a primitive mth root of unity. The field $K = \mathbb{Q}(\omega, \alpha_1, \ldots, \alpha_k)$ is a normal extension of $\mathbb{Q}$. The Čebotarev density theorem says that the density of rational primes that completely split in K is $1/[K : Q]$, and thus nonzero. Let p be such a prime that moreover, does not divide the discriminant of K. Then p splits completely in every subfield of K, in particular in $\mathbb{Q}(\alpha_1)$. From Dirichlet's unit theorem, it follows that the congruence $f(x) \equiv 0 \pmod{p}$ has k distinct solutions, since p does not divide the discriminant of $Q(\alpha_1)$. Let Φ_m be the cyclotomic polynomial of degree m. A similar argument concerning $\mathbb{Q}(\omega)$ shows that the congruence $\Phi_m(x) \equiv 0 \pmod{p}$ has exactly $\phi(m)$ roots. Let r be one of them. Then the order of r in $\mathbb{Z}/p\mathbb{Z}$ is exactly m and thus m divides $p - 1$. Therefore p has the required properties.

Comments 36 *If we required the prime factors to be congruent to 1 modulo 4, then the answer would be positive. In fact, for any integer n, the prime divisors of $4n^2 + 1$ are of the form $4m + 1$.*

6

Algebra and Combinatorics Solutions

6.1 Algebra

Problem 2.1. Set $S_{k,p} = \sum_{i=1}^{p-1} i^k$, for natural numbers p and k. If $p \geq 3$ is prime and $1 < k \leq p - 2$, show that

$$S_{k,p} \equiv 0 \ (\text{mod } p).$$

Solution 2.1. Consider the symmetric polynomial $P(x) = x_1^k + \cdots + x_{p-1}^k$. According to the fundamental theorem of symmetric polynomials, there exists $G \in \mathbb{Z}[x_1, \ldots, x_{p-1}]$ such that $P(x) = G(\sigma_1, \ldots, \sigma_{p-1})$, where $\sigma_i(x_1, \ldots, x_{p-1})$ are the fundamental symmetric polynomials

$$\sigma_i(x_1, \ldots, x_{p-1}) = \sum_{1 \leq j_1 < \cdots < j_i \leq p-1} x_{j_1} \cdots x_{j_i}.$$

Furthermore, the largest n with the property that σ_n actually appears in the development of G satisfies $n \leq \deg P$. In particular, if $k \leq p-2$, then $P = G(\sigma_1, \ldots, \sigma_{p-1})$.

Let us show first that $\sigma_i(1, 2, \ldots, p-1) \equiv 0 \ (\text{mod } p)$. In fact, the theorem of Fermat tells us that $a^{p-1} \equiv 1 \ (\text{mod } p)$, for any number $a \not\equiv 0 \ (\text{mod } p)$. Thus the polynomial $Q = x^{p-1} - 1 \in \mathbb{Z}/p\mathbb{Z}[x]$ with coefficients in $\mathbb{Z}/p\mathbb{Z}$ vanishes for all nonzero classes $\{1 \ldots, p-1\}$ modulo p, and so these are all the roots of Q. Therefore, Q decomposes in $\mathbb{Z}/p\mathbb{Z}[x]$ as $Q = (x - 1) \cdots (x - (p - 1))$. By identifying the coefficients of the right- and left-hand sides, we obtain $\sigma_i(1, \ldots, p-1) \equiv 0 \ (\text{mod } p)$, as claimed. Finally, $S_{k,p} = P(1, \ldots, p-1) = G(0, 0, 0, \ldots, 0) \equiv 0 \ (\text{mod } p)$, which proves the claim.

Observe that if $k = p - 1$, then we have $i^{p-1} \equiv 1 \ (\text{mod } p)$ and hence $S_{p-1,p} \equiv -1 \ (\text{mod } p)$. In particular, we have

$$S_{k,p} = \begin{cases} 0, & \text{if } k \not\equiv -1 \ (\text{mod } p - 1), \\ -1, & \text{if } k \equiv -1 \ (\text{mod } p - 1). \end{cases}$$

Problem 2.2. Let $P = a_0 + \cdots + a_n x^n$ and $Q = b_0 + \cdots + b_m x^m$ be two polynomials with $m \leq n$. Then, $\deg \gcd(P, Q) \geq 1$ if and only if there exist two polynomials K and L such that $\deg K \leq m - 1$, $\deg L \leq n - 1$, and $K \cdot P = L \cdot Q$. Prove that this is equivalent to the vanishing of the following $(n + m) \times (n + m)$ determinant:

$$
\det \begin{pmatrix}
a_0 & a_1 & a_2 & \cdots & a_{m-1} & \cdots & a_{n-1} & a_n & 0 & 0 & \cdots & 0 \\
0 & a_0 & a_1 & \cdots & a_{m-2} & \cdots & a_{n-2} & a_{n-1} & a_n & 0 & \cdots & 0 \\
\vdots & \vdots & \vdots & & \vdots & & \vdots & \vdots & \vdots & \vdots & & \vdots \\
0 & 0 & 0 & \cdots & a_0 & \cdots & a_{n-m-1} & a_{n-m} & a_{n-m+1} & \cdots & & a_n \\
0 & 0 & 0 & \cdots & 0 & \cdots & a_{n-m-2} & a_{n-m-1} & a_{n-m} & & & a_{n-1} \\
\vdots & \vdots & \vdots & & \vdots & & \vdots & \vdots & \vdots & & & \vdots \\
0 & 0 & 0 & \cdots & 0 & 0 & a_0 & a_1 & a_2 & \cdots & & a_m \\
b_0 & b_1 & \cdots & b_{m-1} & b_m & 0 & 0 & 0 & 0 & \cdots & & 0 \\
0 & b_0 & \cdots & b_{m-2} & b_{m-1} & b_m & 0 & 0 & 0 & \cdots & & 0 \\
\vdots & \vdots & & \vdots & \vdots & \vdots & \vdots & \vdots & \vdots & & & \vdots \\
0 & 0 & \cdots & b_0 & b_1 & \cdots & \cdots & \cdots & 0 & \cdots & & 0
\end{pmatrix} = 0.
$$

Solution 2.2. Let us write

$$
P = a_n \prod_{\sum \delta_i = n} (x - x_i)^{\delta_i}, \qquad Q = b_m \prod_{\sum \gamma_i = m} (x - \tilde{x}_i)^{\gamma_i},
$$

$$
L = c_{m-1} \prod_{\sum \xi_i = n-1} (x - y_i)^{\xi_i}, \quad K = d_{m-1} \prod_{\sum \psi_i = m-1} (x - \tilde{y}_i)^{\psi_i},
$$

and then consider the identity $P \cdot K = Q \cdot L$, each side being decomposed into linear terms. According to the pigeonhole principle, after the reduction of the common terms, on each side, left and right, there remains some polynomial of degree $\geq n + m - (n + m - 1) = 1$.

Conversely, if $P = P_0 \cdot P_1$, $Q = P_0 \cdot Q_1$, then $Q_1 \cdot P = P_1 \cdot Q$.

For the second assertion, set $K = \sum u_i x^i$ and $L = \sum w_i x^i$. By identifying the coefficients of $P \cdot K$ and $Q \cdot L$, one obtains the identities

$$
\sum_{j+k=i} a_j u_k = \sum_{j+k=i} b_j w_k.
$$

Alternatively,

$$
D \cdot (u_0, u_1, \ldots, u_{n-1}, -w_0, -w_1, \ldots, -w_{m-1})^\top = 0,
$$

where D denotes the matrix from the claim. Since K and L are nonzero, this implies that $\det D = 0$.

Problem 2.3. Prove that if $P, Q \in \mathbb{R}[x, y]$ are relatively prime polynomials, then the system of equations

$$
\begin{aligned}
P(x, y) &= 0, \\
Q(x, y) &= 0,
\end{aligned}
$$

has only finitely many real solutions.

Solution 2.3. Assume that the value of y is a fixed y_0, so that there exists at least one solution (x, y_0). Then the polynomials $P(x, y_0)$ and $Q(x, y_0)$ from $\mathbb{R}[x]$ have at least one common root, and hence a nontrivial common divisor. According to the preceding problem, this amounts to the vanishing of the determinant of some matrix $D_{y_0} = D(P(x, y_0), Q(x, y_0))$ obtained from the coefficients.

Further, the determinant of D_{y_0} is a nonzero polynomial in y_0. In fact, if D_y were identically zero, then P and Q would have a common divisor, contradicting our assumptions. Therefore the number of such y_0 for which a solution (x, y_0) exists is finite.

In a similar way, the number of those x_0 for which there exist solutions (x_0, y) is finite, by the same argument. This implies that the number of solutions is finite.

Problem 2.4. Let $a, b, c, d \in \mathbb{R}[x]$ be polynomials with real coefficients. Set

$$p = \int_1^x ac \, dt, \quad q = \int_1^x ad \, dt, \quad r = \int_1^x bc \, dt, \quad s = \int_1^x bd \, dt.$$

Prove that $ps - qr$ is divisible by $(x - 1)^4$.

Solution 2.4. Let p, q, r, s be smooth functions such that $p(u) = q(u) = r(u) = s(u) = 0$ and $\det\left(\begin{smallmatrix} p' & q' \\ r' & s' \end{smallmatrix}\right)(u) = 0$. Then the function $\det\left(\begin{smallmatrix} p & q \\ r & s \end{smallmatrix}\right)$, together with its first three derivatives, vanishes at u. The proof is immediate. The given p, q, r, s satisfy the previous conditions at $u = 1$, and hence the claim.

Problem 2.5. 1. Find the minimum number of elements that must be deleted from the set $\{1, \ldots, 2005\}$ such that the set of the remaining elements does not contain two elements together with their product.

2. Does there exist, for any k, an arithmetic progression with k terms in the infinite sequence

$$1, \frac{1}{2}, \ldots, \frac{1}{2005}, \ldots, \frac{1}{n}, \ldots?$$

Solution 2.5. 1. If we extract the numbers $1, 2, 3, \ldots, 44$ and then choose $x, y \in \{45, \ldots, 2005\}$, then $xy \geq 45^2 > 2005$, and thus it cannot belong to the given set. Therefore the number of elements to be deleted is at most 43.

Consider now the triples $(2, 87, 2 \cdot 87), (3, 86, 3 \cdot 86), \ldots, (44, 45, 44 \cdot 45)$ consisting of elements of the given set. We must delete at least one number from each triple in order that the remaining set be free of products. Thus the minimum number of elements to be deleted is 43.

2. Yes, it does. Take for instance $\frac{1}{k!}, \frac{2}{k!}, \ldots, \frac{k}{k!}$.

Problem 2.6. Consider a set S of n elements and $n + 1$ subsets $M_1, \ldots, M_{n+1} \subset S$. Show that there exist $r, s \geq 1$ and disjoint sets of indices $\{i_1, \ldots, i_r\} \cap \{j_1, \ldots, j_s\} = \emptyset$ such that

$$\bigcup_{k=1}^r M_{i_k} = \bigcup_{k=1}^s M_{j_k}.$$

Solution 2.6. Assume that $S = \{a_1, a_2, \ldots, a_n\}$. We associate to the subset M_j the vector $v_j \in \mathbb{R}^n$ having the components $v_j = (v_{1j}, v_{2j}, \ldots, v_{nj})$, given by

$$v_{ij} = \begin{cases} 1, & \text{if } a_i \in M_j, \\ 0, & \text{if } a_i \notin M_j. \end{cases}$$

Since we have $n + 1$ vectors $v_1, \ldots, v_{n+1}$ in $\mathbb{R}^n$, there exists a nontrivial linear combination of them that vanishes. We separate the positive coefficients and the negative coefficients and write down this combination in the form

$$\sum_{i \in I} \lambda_i v_i = \sum_{j \in J} \lambda_j v_j,$$

where I, J are disjoint sets of indices and $\lambda_k > 0$, for $k \in I \cup J$. We claim now that

$$\cup_{i \in I} M_i = \cup_{j \in J} M_j.$$

In fact, assume that $a_k \in \cup_{i \in I} M_i$ and thus $a_k \in M_i$ for some $i \in I$. Then the kth coordinate of the vector v_i is nonzero. All coordinates of the vectors v_s are non-negative and the coefficients λ_i are positive, which implies that the kth coordinate of the vector $\sum_{i \in I} \lambda_i v_i$ is nonzero. Since the kth coordinate of $\sum_{j \in J} \lambda_j v_j$ is nonzero, there should exist some $j \in J$ for which the kth coordinate of v_j is nonzero and hence $a_i \in M_j \subset \cup_{j \in J} M_j$. This proves that $\cup_{i \in I} M_i \subset \cup_{j \in J} M_j$, and by symmetry, we have equality.

Problem 2.7. Let p be a prime number, and $A = \{a_1, \ldots, a_{p-1}\} \subset \mathbb{Z}_+^*$ a set of integers that are not divisible by p. Define the map $f : \mathcal{P}(A) \to \{0, 1, \ldots, p-1\}$ by

$$f(\{a_{i_1}, \ldots, a_{i_k}\}) = \sum_{p-1}^{k} a_{i_p} \ (\text{mod } p), \text{ and } f(\emptyset) = 0.$$

Prove that f is surjective.

Solution 2.7. Let $C_0 = \emptyset$, $C_n = \{a_1, \ldots, a_n\}$, for $n \leq p - 2$. Set $P_n = \{f(\beta) | \beta \subset C_n\}$, and $b_n = f(\{a_n\}) \neq 0$. One obtains easily the following inductive description of the sets P_n:

$$P_{n+1} = \{r + b_{n+1} | r \in P_n \cup \{0\}\} \cup P_n.$$

Furthermore, if $P_n = P_{n+1}$, then $b_{n+1} + r \in P_n$ for all $r \in P_n$. In particular, $b_{n+1} \in P_n$, and by induction, $k b_{n+1} \in P_n$, for any k. Since p is prime and $b_{n+1} \neq 0$, we obtain $P_n = \{0, 1, \ldots, p - 1\}$.

Assume now that $P_n \neq P_{n+1}$ for all n. Since card $P_0 = 1$ and $P_n \subset P_{n+1}$, we derive that card $P_n \geq n + 1$, and hence $P_{p-1} = \mathcal{P}$.

Problem 2.8. Consider the function $F_r = x^r \sin rA + y^r \sin rB + z^r \sin rC$, where $x, y, z \in \mathbb{R}$, $A + B + C = k\pi$, and $r \in \mathbb{Z}_+$. Prove that, if $F_1(x_0, y_0, z_0) = F_2(x_0, y_0, z_0) = 0$, then $F_r(x_0, y_0, z_0) = 0$, for all $r \in \mathbb{Z}_+$.

Solution 2.8. Consider the complex numbers

$$u = x_0(\cos A + i \sin A), \quad v = y_0(\cos B + i \sin B), \quad w = z_0(\cos C + i \sin C).$$

We denote the argument of the complex number z by $\arg z$ and its imaginary part by $\Im z$. We have $\arg u + \arg v + \arg w = k\pi$, $\Im u + \Im v + \Im w = 0$, and $\Im u^2 + \Im v^2 + \Im w^2 = 0$. Now $\Im z = 0$ means that z is real, so that $\Im z^2 = 0$. Thus $2\Im(uv + vw + wu) = \Im((u + v + w)^2) - \Im(u^2 + v^2 + w^2) = 0$. Next, $\arg uvw = \arg u + \arg v + \arg w = k\pi$, and so $\Im uvw = 0$.

Let us consider the polynomial $P_n(x_1, x_2, x_3) = x_1^n + x_2^n + x_3^n$. According to the fundamental theorem of symmetric polynomials, we can write P_n as a polynomial in the fundamental symmetric polynomials:

$$P_n(x_1, x_2, x_3) = G(\sigma_1, \sigma_2, \sigma_3),$$

where $G \in \mathbb{R}[\sigma_1, \sigma_2, \sigma_3]$ is a real polynomial and $\sigma_i, i \in \{1, 2, 3\}$, are the fundamental symmetric polynomials in three variables, namely

$$\sigma_1(x_1, x_2, x_3) = x_1 + x_2 + x_3, \quad \sigma_2(x_1, x_2, x_3) = x_1x_2 + x_2x_3 + x_3x_1,$$
$$\sigma_3(x_1, x_2, x_3) = x_1x_2x_3.$$

Notice now that $\sigma_1(u, v, w) = u + v + w \in \mathbb{R}$, $\sigma_2(u, v, w) = uv + vw + wu \in \mathbb{R}$, $\sigma_3(u, v, w) = uvw \in \mathbb{R}$, which implies that $P_n(u, v, w) \in \mathbb{R}$, and therefore $F_r(x_0, y_0, z_0) = \Im(P_n(u, v, w)) = 0$.

Problem 2.9. Let $T(z) \in \mathbb{Z}[z]$ be a nonzero polynomial with the property that $|T(u_i)| \leq 1$ for all values u_i that are roots of $P(z) = z^n - 1$. Prove that either $T(z)$ is divisible by $P(z)$, or else there exists some $k \in \mathbb{Z}_+, k \leq n - 1$, such that $T(z) \pm z^k$ is divisible by $P(z)$. The same result holds when instead of $P(z)$, we consider $z^n + 1$.

Solution 2.9. Let us write $T(z) = A(z)P(z) + B(z)$, where $\deg B \leq n - 1$ and $A, B \in \mathbb{Z}[x]$. Assume that $T(z)$ is not divisible by $P(z)$, and so $B \neq 0$. Set k for the multiplicity of the root 0 in B, meaning that $B(z) = z^k C(z)$, where $C(0) \neq 0$. We have then $|C(u_i)| = |B(u_i)| = |T(u_i)| \leq 1$.

If we set $Q_d(z) = z^d$, then we can compute easily the sum of roots of unity as

$$\sum_{i=1}^{d} Q_d(u_i) = 0, \text{ if } 1 \leq d \leq n - 1.$$

We want to compute this kind of sum for the polynomial C. Since $\deg C \leq n - 1$, the previous identities show that only the degree-zero part of C has a nontrivial contribution, and thus

$$\sum_{i=1}^{n} C(u_i) = nC(0).$$

This implies that

$$0 < |C(0)| \leq \frac{1}{n} \left| \sum_{i=1}^{n} C(u_i) \right| \leq \frac{1}{n} \sum_{i=1}^{n} |C(u_i)| \leq 1.$$

Now $C(0) \in \mathbb{Z}$ and thus $C(0) = \pm 1$, which implies that either $C(u_i) = 1$, for all i, or else $C(u_i) = -1$, for all i. Since deg $C \leq n - 1$, we derive that either $C(z) = 1$ or else $C(z) = -1$. In particular, $T(z) \pm z^k$ is divisible by $P(z)$.

The case in which $P(z)$ is replaced by $z^n + 1$ is similar.

Problem 2.10.　1. If the map $x \mapsto x^3$ from a group G to itself is an injective group homomorphism, then G is an abelian.
2. If the map $x \mapsto x^3$ from a group G to itself is a surjective group homomorphism, then G is an abelian.
3. Find an abelian group with the property that $x \mapsto x^4$ is an automorphism.
4. What can be said for exponents greater than 4?

Solution 2.10. 1. We have $a^3 b^3 = (ab)^3$, and thus $a^2 b^2 = (ba)^2$, whence $a^4 b^4 = (a^2)^2 (b^2)^2 = (b^2 a^2)^2 = (ab)^4$. Using again that $a^3 b^3 = (ba)^3$, we get $(ab)^3 = (ba)^3$. Since the map is injective, we have $ab = ba$.

2. We continue from above, $a^3 b^3 = (ba)^3 = (ba)^2 ba = a^2 b^2 (ba)$; thus $ab^3 = b^3 a$. Since the homomorphism is surjective, b^3 can take any value in G; hence the relation above shows that G is abelian.

3. Take any nonabelian group of order 8, for instance the dihedral group D_4.

4. For $m > 3$, there exists a nonabelian group G_m such that $x \mapsto x^m$ is the identity. For instance, if p is not of the form $2^n + 1$, then let p be an odd prime factor of $m - 1$ and G_m the group of matrices of the form

$$M = \begin{pmatrix} 1 & a & b \\ 0 & 1 & c \\ 0 & 0 & 1 \end{pmatrix}$$

with arbitrary $a, b, c \in \mathbb{Z}/p\mathbb{Z}$. Then G_m is a nonabelian group of order p^3 such that the order of every element $\neq e$ is p.

Problem 2.11. Let V be a vector space of dimension $n > 0$ over a field of characteristic $p \neq 0$ and let A be an affine map $A : V \to V$. Prove that there exist $u \in V$ and $1 \leq k \leq np$ such that $A^k u = u$.

Solution 2.11. We write $Ax = Bx + a$, and thus $A^k x = B^k x + C_k a$, where $C_k = B^{k-1} + \cdots + B + I$. We use $(B - I)C_k = B^k - I$ to rewrite the equation to solve as $(B - I)C_k u = C_k a$. There exists a solution if and only if $a \in \ker C_k + \text{Im}(B - I)$. We suppose $B - I$ is singular. The restriction of B to the eigenspace S corresponding to the eigenvalue 1 can be written $B_S = I + N$, where N is nilpotent. Therefore

$$C_k | S = \sum_{i=0}^{k-1} (I + N)^i = \sum_{i=0}^{k-1} \sum_{j=0}^{i} C_i^j N^j = \sum_{j=0}^{k-1} \left(\sum_{i=j}^{k-1} C_i^j \right) C_i^j N^j = \sum_{j=0}^{k-1} C_k^{j+1} N^j.$$

If p divides k, then $C_k|S = N^{k-1}$, and thus if $k - 1 \geq \dim S$, then $C_k|S = 0$. Therefore, $S \subset \ker C_k$ if $k = lp \geq 1 + \dim S$.

Since $\text{Im}(B - I) \supset \text{Im}(B - I)^n$, which is complementary to S, we can take $k = (n - 1)p$ if $(n - 1)p \geq \dim S + 1$, in which case $\text{Im}(B - I) + \ker C_k = V$. Otherwise, we have either $n = 1$, or $n = 2$, $p = 2$, and $\dim S = 2$. If $n = 1$, then take $k = p$. If $n = 2 = p$, then take $k = 2$ if $\dim S = 1$ or $B = I$, and $k = 4$ otherwise. In this situation, $k = np$ is convenient, while in the former case $k = (n-1)p$ suffices. This proves the claim.

Problem 2.12. Find the cubic equation the zeros of which are the cubes of the roots of the equation $x^3 + ax^2 + bx + c = 0$.

Solution 2.12. It is known that if the matrix M has eigenvalues $\lambda_1, \ldots, \lambda_n$ and if F is a polynomial, then $F(M)$ has eigenvalues $F(\lambda_1), \ldots, F(\lambda_n)$. We consider now the matrix

$$M = \begin{pmatrix} 0 & 0 & -c \\ 1 & 0 & -b \\ 0 & 1 & -a \end{pmatrix},$$

which has the characteristic polynomial $x^3 + ax^2 + bx + c$. The characteristic polynomial of M^3 will then have as roots the cubes of the given polynomial. We now compute easily the characteristic polynomial of M^3 as $x^3 + (a^3 - 3ab + c)x^2 + (b^3 - 3ab + 3c^2)x + c^3 = 0$.

Problem 2.13. Assume that the polynomials $P, Q \in \mathbb{C}[x]$ have the same roots, possibly with different multiplicities. Suppose, moreover, that the same holds true for the pair $P + 1$ and $Q + 1$. Prove that $P = Q$.

Solution 2.13. Let A and B be the sets of roots of P and $P + 1$, respectively. Suppose that $\deg P = m \geq n = \deg Q$. Obviously, A and B should be disjoint. It follows that P' has at least $m - r$ roots from the set A and at least $m - s$ roots from the set B, because multiple roots of P are roots of P'. Since $\deg P' = m - 1$, we derive that $(m - r) + (m - s) \leq m - 1$ and thus $r + s > m$. Now, the union $A \cup B$ has $r + s$ elements, because these sets are disjoint. On the other hand, each element of $A \cup B$ is a root of the polynomial $P - Q$, of degree m. Thus $P - Q$ has at least $m + 1$ roots, and thus it vanishes identically.

Problem 2.14. Determine $r \in \mathbb{Q}$, for which $1, \cos 2\pi r, \sin 2\pi r$ are linearly dependent over $\mathbb{Q}$.

Solution 2.14. We consider $0 \leq r < 1$. We suppose that there exists a linear relation $a + b \cos 2\pi r + c \sin 2\pi r = 0$, with $a, b, c, r \in \mathbb{Q}$. This implies that

$$(b^2 + c^2) \cos^2 2\pi r + 2ab \cos \pi r + a^2 - c^2 = 0,$$

i.e., $\cos 2\pi r$ satisfies a quadratic equation with rational coefficients. The same holds for $\sin 2\pi r$.

Consider now $\xi = \cos 2\pi r + i \sin 2\pi r$. If n is the smaller denominator of r, then ξ is a primitive root of nth order.

On the other hand, since both $\cos 2\pi r$ and $\sin 2\pi r$ satisfy quadratic equations, then ξ satisfies an equation of order 4, with rational coefficients.

We know that $\xi^n = 1$. Further, the irreducible polynomial of smallest degree that divides $x^n - 1$ is $1 + x + \cdots + x^{\varphi(n)}$, of degree $\varphi(n)$, where $\varphi(n)$ is Euler's totient function.

Our previous discussion implies that $\varphi(n)$ must be a divisor of 4, and hence $\varphi(n) \in \{1, 2, 4\}$. Therefore, we have the following cases:

1. $\varphi(n) = 1$ and hence $n \in \{1, 2\}$, and so $r \in \left\{0, \frac{1}{2}\right\}$.

2. $\varphi(n) = 2$, and thus $n \in \{3, 4, 6\}$, whence $r \in \left\{\frac{1}{4}, \frac{3}{4}, \frac{1}{3}, \frac{2}{3}, \frac{1}{6}, \frac{5}{6}\right\}$.

3. $\varphi(n) = 4$, and thus $n \in \{5, 8, 10, 12\}$, whence $r \in \left\{\frac{1}{8}, \frac{3}{8}, \frac{5}{8}, \frac{7}{8}, \frac{1}{12}, \frac{5}{12}, \frac{7}{12}, \frac{11}{12}\right\}$.

We omitted in the list above those r having denominator 5 or 10 because $\sin \frac{2\pi}{5} = \sqrt{\frac{1}{2}(5 + \sqrt{5})}$ does not satisfy any quadratic equation with rational coefficients. Therefore, $r \in \left\{x \,|\, x = \frac{m}{24},\ \gcd(m, 24) \neq 1, m \in \mathbb{Z}\right\}$.

Problem 2.15. 1. Prove that there exist $a, b, c \in \mathbb{Z}$, not all zero, such that $|a|, |b|, |c| < 10^6$ $\left|a + b\sqrt{2} + c\sqrt{3}\right| < 10^{-11}$.

2. Prove that if $0 \leq |a|, |b|, |c| < 10^6$, $a, b, c \in \mathbb{Z}$, and at least one of them is nonzero, then $\left|a + b\sqrt{2} + c\sqrt{3}\right| > 10^{-21}$.

Solution 2.15. 1. Consider the set $S = \left\{r + s\sqrt{2} + t\sqrt{3}\right\}_{r,s,t \in \{0,1,\ldots,10^6-1\}}$. Then card $S = 10^{18}$. Set $d = \left(1 + \sqrt{2} + \sqrt{3}\right)10^6$. Then all elements $x \in S$ are bounded by d, i.e., $0 \leq x < d$. One divides the interval $[0, d]$ into $10^{18} - 1$ equal intervals $[(k-1)e, ke]$, where $e = 10^{-18}d$ and $k = 1, \ldots, 10^{18} - 1$. By the pigeonhole principle, there exist two elements $x, y \in S$ lying in the same interval, and hence satisfying

$$|x - y| \leq e < 10^{-11}$$

2. Let $F_1 = a + b\sqrt{2} + c\sqrt{3}$, $F_2 = a + b\sqrt{2} - c\sqrt{3}$, $F_3 = a - b\sqrt{2} + c\sqrt{3}$, $F_4 = -a + b\sqrt{2} + c\sqrt{3}$. Now, $1, \sqrt{2}, \sqrt{3}$ are linearly independent over $\mathbb{Z}$, and thus $F_i \neq 0$, for all $1 \leq i \leq 4$. Further, $F = F_1 F_2 F_3 F_4 \in \mathbb{Z}$, since the product of conjugates contains no more square roots. This implies that $|F_1| \geq 1/|F_2 F_3 F_4| > 10^{-21}$, because $1/|F_i| > 10^{-7}$.

Problem 2.16. Prove that if $n > 2$, then we do not have any nontrivial solutions for the equation

$$x^n + y^n = z^n,$$

where x, y, z are rational functions. Solutions of the form $x = af$, $y = bf$, $z = cf$, where f is a rational function and a, b, c are complex numbers satisfying $a^n + b^n = c^n$, are called *trivial*.

Solution 2.16. Any rational solution yields a polynomial solution, by clearing the denominators. Let further (f, g, h) be a polynomial solution for which the degree $r = \max\{\deg(f), \deg(g), \deg(h)\}$ is minimal among all polynomial solutions, and where $r > 0$.

Consider the root of unity $\xi = \exp\left(\frac{2\pi i}{n}\right)$ and assume that f, g, and h are relatively prime. The equation can be written as

$$\prod_{j=0}^{n-1}(f - \xi^j h) = g^n.$$

Now, $\gcd(f, h)$ and $\gcd(f, g)$ are relatively prime, so that $f - \xi^j h$ and $f - \xi^k h$ are relatively prime when $k \neq j$. Since the factorization of polynomials is unique, we must have $g = g_1 \cdots g_{n-1}$, where g_j are polynomials satisfying $g_j^n = f - \xi^j h$.

Consider now the set $\{f - h, f - \xi h, f - \xi^2 h\}$. Since $n > 2$, these elements belong to the two-dimensional space generated by f and h over $\mathbb{C}$. Thus there exists a vanishing linear combination with complex coefficients in these three elements. Thus, there exist $a_i \in \mathbb{C}$ so that $a_0 g_0^n + a_1 g_1^n = a_2 g_2^n$. We then set $h_j = \sqrt[n]{a_j} g_j$, and observe that

$$h_0^n + h_1^n = h_2^n.$$

Moreover, the polynomials h_i and h_j are relatively prime if $i \neq j$ and $\max \deg(h_i) < r$, which contradicts our choice of r. This proves the claim.

Problem 2.17. A table is an $n \times k$ rectangular grid drawn on the torus, every box being assigned an element from $\mathbb{Z}/2\mathbb{Z}$. We define a transformation acting on tables as follows. We replace all elements of the grid simultaneously, each element being changed into the sum of the numbers previously assigned to its neighboring boxes. Prove that iterating this transformation sufficiently many times, we always obtain the trivial table filled with zeros, no matter what the initial table was, if and only if $n = 2^p$ and $k = 2^q$, for some integers p, q. In this case, we say that the respective $n \times k$ grid is nilpotent.

Solution 2.17. Consider the $m \times m$ square matrices $(m \geq 2)$

$$D_m = \begin{pmatrix} 0 & 1 & 0 & 0 & \cdots & 0 & 1 \\ 1 & 0 & 1 & 0 & \cdots & 0 & 0 \\ 0 & 1 & 0 & 1 & \cdots & 0 & 0 \\ \vdots & \vdots & \vdots & \vdots & & \vdots & \vdots \\ 0 & 0 & 0 & 0 & \cdots & 0 & 1 \\ 1 & 0 & 0 & 0 & \cdots & 1 & 0 \end{pmatrix}.$$

We will write $\equiv$ below if the equality holds (only) for matrices of entries modulo 2. Then the transformation T from the statement acts on the vector space $M_{n,k}(\mathbb{Z}/2\mathbb{Z})$ of $n \times k$ matrices with entries in $\mathbb{Z}/2\mathbb{Z}$ as follows:

$$T(X) = D_n X + X D_k, \quad \text{for } X \in M_{n,k}(\mathbb{Z}/2\mathbb{Z}).$$

By induction on s, we have that

$$T^{2^s}(X) \equiv D_n^{2^s} X + X D_k^{2^s}.$$

Thus there exists N such that $T^N(X) \equiv 0$ if and only if there exists some s such that $D_n^{2^s} X \equiv X D_k^{2^s}$.

Moreover, we can prove easily that for two matrices A and B of appropriate sizes, we have $AX \equiv XB$ for all $X \in M_{n,k}(\mathbb{Z}/2\mathbb{Z})$ if and only if A and B are multiples of the identity by the same scalar. This follows, for instance, if we take X having all but one column (or line) trivial. This assertion is a special case of Schur's lemma.

Thus T is nilpotent if and only if there exist s and $\beta \in \mathbb{Z}/2\mathbb{Z}$ such that $D_n^{2^s} = \beta \cdot \mathbf{1}_n$ and $D_k^{2^s} = \beta \cdot \mathbf{1}_k$. Thus the $n \times k$ grid is nilpotent if and only if the $n \times 1$ and $1 \times k$ grids are nilpotent.

Let us analyze the case of the $n \times 1$ grid. Set $\mathbf{a} = (a_1, \ldots, a_n)$ and $\mathbf{a}^{[k]} = D_n^k \mathbf{a} = (a_1^{[k]}, \ldots, a_n^{[k]})$. By induction, we have

$$a_s^{[2^k]} \equiv a_{s+2^{k-1}} + a_{s-2^{k-1}},$$

where the indices are modulo n. In particular, if $n = 2^q$, then $D_n^{2^q} = 0$ and the grid is nilpotent. This proves the "if" part of the statement.

Assume next that n is odd and there exists some r such that $\mathbf{a}^{[r]} \equiv 0$, for any $\mathbf{a}$. Take such an r that is minimal. If $n \geq 3$, then $r \geq 3$. We claim now that $a_s^{[r-1]} \equiv 1$ for all s. This is true at least for one s, since r is minimal. Since $a_s^{[r]} \equiv a_{s+1}^{[r-1]} + a_{s-1}^{[r-1]}$, we find that the terms corresponding to even indices (modulo n) are all equal. But n is odd and thus all terms are equal. Further, we have $a_s^{[r-1]} = a_{s-1}^{[r-2]} + a_{s+1}^{[r-2]}$. We obtain then that

$$n \equiv \sum_{s=1}^{n} a_s^{[r-1]} = \sum_{s=1}^{n} (a_{s-1}^{[r-2]} + a_{s+1}^{[r-2]}) = 2 \sum_{s=1}^{n} a_s^{[r-2]} \equiv 0 \pmod{2},$$

which is a contradiction. This proves actually that $\mathbf{a}^{[r]} \equiv 0$ when n is odd if and only if $\mathbf{a} \in \{(0, 0, \ldots, 0), (1, 1, \ldots, 1)\}$.

Finally, consider $n = 2^m h$, with h odd. For a vector a of length n, we set

$$\Phi_s(\mathbf{a}) = (a_s, a_{s+2^m}, \ldots, a_{s+(h-1)2^m}).$$

We observe now that

$$\Phi_s(T^{2^m}(a)) = T(\Phi_s(\mathbf{a}))$$

because $a_s^{[2^{m+1}]} = a_s + a_{s+2^m}$. Since the vector $\Phi_s(\mathbf{a})$ is of length h (which is odd), the previous claim shows us that $T^N(\Phi_s(\mathbf{a})) \equiv 0$ for some N only if $\Phi_s(\mathbf{a}) \in \{(0, 0, \ldots, 0), (1, 1, \ldots, 1)\}$. Moreover, this happens for all values of s. This condition is trivial only when $h = 1$. This proves the "only if" part.

Comments 37 *This result was stated in the 1980s as an open question in the journal Kvant. The same method can be used to show that the $n \times k$ grids in the plane are* *nilpotent if and only if* $n = 2^p - 1$ *and* $k = 2^q - 1$ *for some natural numbers* p, q. *A nice corollary is the following. Define the sequence* (α_k) *by the recurrence relations*

$$\alpha_1 = 0, \quad \alpha_{k+1} = \min(2\alpha_k + 2, 2N - 2\alpha_k - 1).$$

Then there exists some k *such that* $\alpha_k = N - 1$ *if and only if the parameter* N *has the form* $2^q - 1$ *for some natural number* q. *The solution given here is from:*

- V. Boju, L. Funar: *Iterative processes for* $\mathbb{Z}_2^n$, Analele Univ. Craiova 15 (1987), 33–38.

6.2 Algebraic Combinatorics

Problem 2.18. Let us consider a four-digit number N whose digits are not all equal. We first arrange its digits in increasing order, then in decreasing order, and finally, we subtract the two obtained numbers. Let $T(N)$ denote the positive difference thus obtained. Show that after finitely many iterations of the transformation T, we obtain 6174.

Solution 2.18. Let N have the digits a, b, c, d.

1. Assume that $a \geq b = c \geq d$. Then in $T(N)$, the sum of the first and the fourth digits is 9, as well as the sum of the second and the third. These give five combinations of four digits that give 6174, by applying T once more.

2. Otherwise, $a \geq b > c \geq d$. Then in $T(N)$, the sum of the digits placed at extremities is 10, while the sum of the middle ones is 8. This gives 25 combinations all leading to the desired result, namely 6174.

Comments 38 *Let* a *be a positive integer having* r *digits, not all equal, in base* g. *Let* a' *be the* g-*adic integer formed by arranging the digits of* a *in descending order and let* a'' *be that formed by ascending order. Define* $T(a) = a' - a''$. *Then* a *is said to be self-producing if* $T(a) = a$. *There exists an algorithm for obtaining self-producing numbers in any base.*

Such a fixed point k *of* T *is called a* (g-*adic*) *Kaprekar constant if it has the further property that every* r-*digit integer* a *eventually yields* k *on repeated iteration of* T *and moreover,* k *is fixed by* T. *D.R. Kaprekar found that* 6174 *has this nice property in a short note published in 1949. Ludington proved that for large* $r \geq n_g$ *there does not exist a Kaprekar constant. Here* n_g *is given by*

$$n_g = \begin{cases} g - 1 + (k-1)\delta, & \text{if } g = 2k, \\ 2k + 1 + (g - k - 2)\delta, & \text{if } g = 2^t k + 2^{t-1} + 1, \\ 3k - 4 + (g-1)\delta, & \text{if } g = 2^t + 1, \end{cases}$$

and $\delta \in \{0, 1\}$ *is equal to* g (mod 2). *Further improvements by Lapenta, Ludington, and Prichett showed that there is no Kaprekar constant in base* 10 *when the number of digits is* $r > 4$.

Hasse and Prichett found that there exists a 4-digit Kaprekar constant in base g if and only if $g = 2^n \cdot 5$, where n is either 0 or an odd number.

Moreover, the situation is completely understood for 3-digit numbers. If the base g is odd, then there is a Kaprekar constant given by $\left(\frac{r-2}{2}, r - 1, \frac{r}{2}\right)$, which is reached within $\frac{r+2}{2}$ iterations of T (respectively 1, if $r = 2$). For instance, 495 is a Kaprekar constant if $g = 10$. If the base g is even, then there is no Kaprekar constant. More precisely, iterations of T will eventually reach the loop of length 2 formed by the pair of numbers $\left(\frac{r-3}{2}, r - 1, \frac{r+1}{2}\right)$, $\left(\frac{r-1}{2}, r - 1, \frac{r-1}{2}\right)$, after at most $\frac{r+1}{2}$ steps (respectively 1 if $r = 3$).

- K.E. Eldridge and S. Sagong: *The determination of Kaprekar convergence and loop convergence of all three-digit numbers,* Amer. Math. Monthly 95 (1988), 105–112.
- J.F. Lapenta, A.L. Ludington, and G.D. Prichett: *An algorithm to determine self-producing r-digit g-adic integers,* J. Reine Angew. Math. 310 (1979), 100–110.
- G.D. Prichett, A.L. Ludington, and J.F. Lapenta: *The determination of all decadic Kaprekar constants,* Fibonacci Quart. 19 (1981), 45–52.
- A.L. Ludington: *A bound on Kaprekar constants,* J. Reine Angew. Math. 310 (1979), 196–203.
- H. Hasse and G.D. Prichett: *The determination of all four-digit Kaprekar constants,* J. Reine Angew. Math. 299/300 (1978), 113–124.

Problem 2.19. Find an example of a sequence of natural numbers $1 \le a_1 < a_2 < \cdots < a_n < a_{n+1} < \cdots$ with the property that every $m \in \mathbb{Z}_+$ can be uniquely written as $m = a_i - a_j$, for $i, j \in \mathbb{Z}_+$.

Solution 2.19. We consider the sequence

$$a_1 = 1, \quad a_2 = 2,$$
$$a_{2n+1} = 2a_{2n},$$
$$a_{2n+2} = a_{2n+1} + r_n,$$

where r_n is the smallest natural number that cannot be written in the form $a_i - a_j$, with $i, j \le 2n + 1$.

Comments 39 *One does not know the minimal growth of such a sequence a_k.*

Problem 2.20. Consider the set of $2n$ integers $\{\pm a_1, \pm a_2, \ldots, \pm a_n\}$ and $m < 2^n$. Show that we can choose a subset S such that

1. The two numbers $\pm a_i$ are not both in S;
2. The sum of all elements of S is divisible by m.

Solution 2.20. Let $S_1, \ldots, S_{2^n-1}$ be the $2^n - 1$ nonempty distinct subsets of the set $\{a_1, \ldots, a_n\}$, where $a_i \ge 0$. Let $F(S_i)$ denote the sum of the elements of S_i. Then, by the pigeonhole principle, there exist i, j such that $F(S_i) \equiv F(S_j) \pmod{m}$. Consider next the set $S = \{S_i \setminus S_j\} \cup -\{S_j \setminus S_i\}$. We derive that $F(S) \equiv 0 \pmod{m}$.

Problem 2.21. Show that for every natural number n there exist prime numbers p and q such that n divides their difference.

Solution 2.21. Consider the following arithmetic progression: $1, 1+n, 1+2n, \ldots, 1+rn, \ldots$. According to Dirichlet's theorem, there exist infinitely many prime numbers among the terms of this progression. Let p, q be two of these prime numbers; then $p = 1 + nr$ and $q = 1 + ns$, where $r \neq s$. This yields $n(r - s) = p - q$, as claimed.

An alternative solution is to consider the set of n arithmetic progressions of ratio n starting respectively at $0, 1, 2, \ldots, n - 1$. Since the set of prime numbers is infinite (this is elementary), there exists at least one progression having infinitely many prime numbers among its terms. The argument above settles the claim.

Problem 2.22. An even number, $2n$, of knights arrive at King Arthur's court, each one of them having at most $n - 1$ enemies. Prove that Merlin the wizard can assign places for them at a round table in such a way that every knight is sitting only next to friends.

Solution 2.22. 1. We consider the friendship graph G defined below: its vertices are in bijection with the knights and the edges are joining pairs of vertices whose respective knights are not enemies. The degree of each vertex is at least n. According to Dirac's theorem, such a graph admits a Hamiltonian cycle, and this yields the wanted assignment of places.

2. Choose an arbitrary assignment of places in which we have two neighbors, A and B, who are enemies. Let us assume that A lies on the right-hand side of B. According to the pigeonhole principle, there exists another pair of enemies, say $\tilde{A}, \tilde{B}$, who are neighbors, and moreover, $\tilde{A}$ is on the right of $\tilde{B}$.

Let us switch all the places starting at A (and lying on the right side of A) and ending at $\tilde{B}$, using a symmetry. After such a transformation, the number of enemy pairs (A, B) is diminished by 2. Applying such transforms iteratively, one obtains a position in which all neighbors are friends.

Problem 2.23. Let $r, s \in \mathbb{Z}_+$. Find the number of 4-tuples of positive integers (a, b, c, d) that satisfy $3^r 7^s = \mathrm{lcm}(a, b, c) = \mathrm{lcm}(a, b, d) = \mathrm{lcm}(a, c, d) = \mathrm{lcm}(b, c, d)$.

Solution 2.23. The numbers a, b, c, d are of the form $3^{m_i} 7^{n_i}$, $1 \leq i \leq 4$, where $0 \leq m_i \leq r$ and $0 \leq n_i \leq s$. Also, $m_i = r$ for at least two values of i, and $n_i = s$ for at least two values of i. We have then:

   (1) one possibility that $m_i = r$ for all four i;
   (2) $4r$ possibilities that precisely one $m_i \in \{0, \ldots, r - 1\}$;
   (3) $C_4^2 r^2$ possibilities that exactly two $m_i \in \{0, \ldots, r - 1\}$.

Therefore, there are $\left(1 + 4r + 6r^2\right)$ possibilities for the m_i's and a similar number for the n_i's, yielding a total of $\left(1 + 4r + 6r^2\right)\left(1 + 4s + 6s^2\right)$ possibilities.

Problem 2.24. 1. Let $n \in \mathbb{Z}_+$ and p be a prime number. Denote by $N(n, p)$ the number of binomial coefficients C_n^s that are not divisible by p. Assume that n is written in base p as $n = n_0 + n_1 p + \cdots + n_m p^m$, where $0 \leq n_j < p$, for all $j \in \{0, 1, \ldots, m\}$. Prove that $N(n, p) = (n_0 + 1)(n_1 + 1) \cdots (n_m + 1)$.

2. Write k in base p as $k = k_0 + k_1 p + \cdots + k_s p^s$, with $0 \le k_j \le p - 1$, for all $j \in \{0, 1, \ldots, s\}$. Prove that

$$C_n^k \equiv C_{n_0}^{k_0} C_{n_1}^{k_1} \cdots C_{n_s}^{k_s} \pmod{p}.$$

Solution 2.24. 1. It is clear that $C_p^k \equiv 0 \pmod{p}$ for all $k \in \{1, 2, \ldots, p\}$. Thus $(1 + x)^p \equiv 1 + x^p$ in $(\mathbb{Z}/p\mathbb{Z})[x]$. By induction, we find that $(1 + x)^{p^n} \equiv 1 + x^{p^n}$ in $(\mathbb{Z}/p\mathbb{Z})[x]$ for any natural number n. We have then the following congruences in $(\mathbb{Z}/p\mathbb{Z})[x]$:

$$(1+x)^n = (1+x)^{n_0}(1+x)^{n_1 p} \cdots (1+x)^{n_m p^m} \equiv (1+x)^{n_0}(1+x^p)^{n_1} \cdots (1+x^{p^m})^{n_m}.$$

When developing factors on the right-hand side, we obtain a sum of factors $x^{a_0 + a_1 p + a_2 p^2 + \cdots + a_m p^m}$, with a coefficient that is nonzero modulo p. Since every number can be uniquely written in base p, all these factors are distinct and their coefficients modulo p are nonzero. There are exactly $(n_0 + 1)(n_1 + 1) \cdots (n_m + 1)$ such factors, and therefore as many binomials not divisible by p.

2. As observed above, the factor x^k appears in the form $x^{a_0 + a_1 p + a_2 p^2 + \cdots + a_m p^m}$; since k can be uniquely written in base p, we have $a_i = k_i$. This implies that the coefficient of x^k is $C_{n_0}^{k_0} C_{n_1}^{k_1} \cdots C_{n_j}^{k_j}$, and hence the claim.

Problem 2.25. Define the sequence T_n by $T_1 = 2$, $T_{n+1} = T_n^2 - T_n + 1$, for $n \ge 1$. Prove that if $m \ne n$, then T_m and T_n are relatively prime, and further, that

$$\sum_{i=1}^{\infty} \frac{1}{T_i} = 1.$$

Solution 2.25. We prove by induction that $T_{n+1} = 1 + T_1 T_2 \cdots T_n$. In fact, we have

$$T_{n+1} = T_n^2 - T_n + 1 = T_n(T_n - 1) + 1 = T_n(T_{n-1} T_{n-2} \cdots T_1) + 1 = 1 + T_1 \cdots T_n.$$

Then take $m < n$. Therefore, T_m divides $T_1 \cdots T_{n-1} = T_n - 1$ and thus $\gcd(T_m, T_n) = 1$. Further, we prove by induction that

$$\sum_{i=1}^{n} \frac{1}{T_i} = 1 - \frac{1}{T_{n+1} - 1}.$$

In fact,

$$\sum_{i=1}^{k+1} \frac{1}{T_i} = 1 - \frac{1}{T_{k+1} - 1} + \frac{1}{T_{k+1}} = 1 - \frac{1}{T_{k+1}(T_{k+1} - 1)} = 1 - \frac{1}{T_{k+2} - 1}.$$

Since T_n tends to infinity, we therefore obtain $\sum_{i=1}^{\infty} \frac{1}{T_i} = 1$.

Problem 2.26. Let $\alpha, \beta > 0$ and consider the sequences

$$[\alpha], [2\alpha], \ldots, [k\alpha], \ldots; [\beta], [2\beta], \ldots, [k\beta], \ldots,$$

where the brackets denote the integer part. Prove that these two sequences taken together enumerate $\mathbb{Z}_+$ in an injective manner if and only if

$$\alpha, \beta \in \mathbb{R} \setminus \mathbb{Q} \text{ and } \frac{1}{\alpha} + \frac{1}{\beta} = 1.$$

Solution 2.26. Set $A_N = \{1, 2, \ldots, N\}$; take k maximal such that $k\alpha < N + 1$ and l maximal such that $l\beta < N + 1$. Therefore, the following inequalities hold: $\left[\frac{N}{\alpha}\right] \le k \le \left[\frac{N+1}{\alpha}\right]$ and $\left[\frac{N}{\beta}\right] \le l \le \left[\frac{N+1}{\beta}\right]$.

Since A_N is injectively enumerated by the two sequences, we have $k + l = N$, and hence

$$\left[\frac{N}{\alpha}\right] + \left[\frac{N}{\beta}\right] \le N \le \left[\frac{N+1}{\alpha}\right] + \left[\frac{N+1}{\beta}\right].$$

Letting N go to infinity, we obtain $\frac{1}{\alpha} + \frac{1}{\beta} = 1$.

If $\alpha \in \mathbb{Q}$, then also $\beta \in \mathbb{Q}$. In this case, write $\alpha = \frac{m}{n}$ and $\beta = \frac{p}{q}$. It follows that $[\alpha n q p] = [\beta m q n]$, which contradicts the injectivity assumption. Therefore $\alpha \notin \mathbb{Q}$ and $\beta \notin \mathbb{Q}$.

Conversely, we have

$$N+1 = \frac{N+1}{\alpha} + \frac{N+1}{\beta} > \left[\frac{N+1}{\alpha}\right] + \left[\frac{N+1}{\beta}\right] > \frac{N+1}{\alpha} + \frac{N+1}{\beta} - 2 = N - 1,$$

whence

$$\left[\frac{N+1}{\alpha}\right] + \left[\frac{N+1}{\beta}\right] = N.$$

In particular, using the notation from above, we have $k + l = N$.

If the two sequences enumerate A_N in a surjective manner, then they will also enumerate A_N injectively, because $k+l = N$. It suffices then to prove the surjectivity.

Let us assume the contrary. Then there exist $u, x, y \in \mathbb{Z}_+$ such that $x\alpha < u < u + 1 < (x + 1)\alpha$ and $y\beta < u < u + 1 < (y + 1)\beta$. Dividing by α, β respectively and adding up the inequalities, we obtain

$$x + y < \frac{u}{\alpha} + \frac{u}{\beta} < \frac{u + 1}{\alpha} + \frac{u + 1}{\beta} < x + 1 + y + 1,$$

which amounts to

$$x + y < u < u + 1 < x + 2 + y.$$

This is false, because x, y, u are integers, and our claim follows.

Problem 2.27. We say that the sets $S_1, S_2, \ldots, S_m$ form a complementary system if they make a partition of $\mathbb{Z}_+$, i.e., every positive integer belongs to a unique set S_i. Let $m > 1$ and $\alpha_1, \ldots, \alpha_m \in \mathbb{R}_+$. Then the sets

$$S_i = \{[n\alpha_i], \text{ where } n \in \mathbb{Z}_+\}$$

form a complementary system only if

$$m = 2, \quad \alpha_1^{-1} + \alpha_2^{-1} = 1, \quad \text{and } \alpha_1 \in \mathbb{R} \setminus \mathbb{Q}.$$

Solution 2.27. If all $\alpha_j < 1$, then the collection S_j contains twice some number. Let $1 \in S_1$, and thus $[\alpha_1] = 1$, $\alpha_1 \neq 1$. We prove first that $S_1 = \{[n\alpha_1]; n \in \mathbb{Z}_+\}$ and $T = \{[n\beta - \epsilon]; n \in \mathbb{Z}_+\}$ form a complementary system, where

$$\beta = \frac{\alpha_1}{\alpha_1 - 1}, \quad \text{and } \epsilon = \begin{cases} (2(a - c))^{-1}, & \text{if } \alpha_1 = \frac{a}{c} \in \mathbb{Q}, \\ 0, & \text{otherwise.} \end{cases}$$

Let $\Xi = \{n\alpha_1, n\beta - \epsilon; n \in \mathbb{Z}_+\}$. It is sufficient to show that for any integer $M > 1$, we have the formula

$$N = \operatorname{card}(\{x \in \Xi; x < M\}) = M - 1.$$

Now, $n\alpha_1 < M$ is equivalent to $n\alpha_1 + \delta < M$, where $\delta = (2c)^{-1}$ if $\alpha_1 = a/c \in \mathbb{Q}$, while $\delta = 0$, if $\alpha \in \mathbb{R} \setminus \mathbb{Q}$. The maximum value for n is therefore $\left[\frac{M-\delta}{\alpha_1}\right]$. In a similar way, the number of elements of the form $n\beta - \epsilon$ that are less than M is $\left[\frac{M+\epsilon}{\beta}\right]$. This implies that $N = [(M - \delta)/\alpha_1] + [(M + \epsilon)/\beta]$. Now we have

$$\frac{M - \delta}{\alpha_1} - 1 < \left[\frac{M - \delta}{\alpha_1}\right] < \frac{M - \delta}{\alpha_1},$$

$$\frac{M + \epsilon}{\beta} - 1 < \left[\frac{M + \epsilon}{\beta}\right] < \frac{M + \epsilon}{\beta},$$

and by summing up the two inequalities, we obtain $M - 2 < N < M$, and therefore $N = M - 1$, as claimed.

Now let k be the smallest integer $k \notin S_1$. Then there exists some α_2 such that $k = [\alpha_2] = [\beta - \epsilon] \geq 2$. We have $n\beta - 1 \leq [n\beta - \epsilon] < n\beta - \epsilon$, and therefore

$$[(n + 1)\beta - \epsilon] - [n\beta - \epsilon] < (n + 1)\beta - \epsilon - n\beta + 1 = \beta + 1 - \epsilon \leq k + 2 - \epsilon,$$
$$[(n + 1)\beta - \epsilon] - [n\beta - \epsilon] > (n + 1)\beta - 1 - n\beta + \epsilon = \beta - 1 + \epsilon > k - 1 + 2\epsilon.$$

Therefore

$$k \leq [(n + 1)\beta - \epsilon] - [n\beta - \epsilon] = [(n + 1)\alpha_2] - [n\alpha_2] \leq k + 1.$$

The difference between two consecutive terms of the sequence $[n\beta - \epsilon]$ is equal to the difference between consecutive missing terms from the sequence $[n\alpha_1]$; that is, k or $k + 1$ is the same as the difference between two consecutive terms of the sequence $[n\alpha_2]$. This implies the fact that the jth term that does not belong to S_1 is precisely $[j\alpha_2] = [j\beta - \epsilon]$. This implies that $m = 2$.

If $\alpha_1 \in \mathbb{Q}$, then $\alpha_2 = b/d$, and for $j = d$, we obtain $[j\alpha_2] > [j\beta - \epsilon]$, which is false.

Comments 40 *The fact that the two sequences from the previous problem are complementary is known as the Rayleigh–Beatty theorem, since S. Beatty proposed it as a problem in Amer. Math. Monthly in 1926. Presumably, this was known to Rayleigh, who mentioned it without proof in 1894. In 1927, J.V. Uspensky proved that if the m sequences are complementary, then $m \leq 2$; his proof was simplified in several papers by Skolem, Graham, and Fraenkel, who also provided a far-reaching generalization in 1969.*

- S. Beatty: *Problem 3173,* Amer. Math. Monthly 33 (1926), 3, 156.
- J. Lambek, L. Moser: *Inverse and complementary sequences of natural numbers,* Amer. Math. Monthly 61 (1954), 454–458.
- A.S. Fraenkel: *Complementary systems of integers,* Amer. Math. Monthly 84 (1977), 114–115.
- A.S. Fraenkel: *The bracket function and complementary sets of integers,* Canad. J. Math. 21 (1969), 6–27.
- R.L. Graham: *On a theorem of Uspensky,* Amer. Math. Monthly 70 (1963), 407–409.
- Th. Skolem: *On certain distributions of integers in pairs with given differences,* Math. Scand. 5 (1957), 57–68.
- J.V. Uspensky, M.A. Heaslet: *Elementary Number Theory,* McGraw-Hill, New York, 1939.

Problem 2.28. Let $f : \mathbb{Z}_+ \to \mathbb{Z}_+$ be an increasing function and set

$$F(n) = f(n) + n, \ \ G(n) = f^*(n) + n,$$

where $f^*(n) = \mathrm{card}(\{x \in \mathbb{Z}_+; 0 \leq f(x) < n\})$. Then $\{F(n); n \in \mathbb{Z}_+\}$ and $\{G(n); n \in \mathbb{Z}_+\}$ are complementary sequences. Conversely, any two complementary sequences can be obtained this way using some nondecreasing function f.

Solution 2.28. We compute the number N of integers from $\{F(n), G(n), n \in \mathbb{Z}_+\}$ that are smaller than M. Suppose that k terms of the form $G(n)$ are smaller than M. Then

$$f^*(k) + k < M \leq f^*(k+1) + k + 1,$$

and so $f^*(k) < M - k < f^*(k+1) + 1$. Since $f^*(k) = \mathrm{card}\{x \in \mathbb{Z}_+; f(x) < k\} < M - k$, we derive $f(M - k) \geq k$. Further, $f^*(k+1) = \{x \in \mathbb{Z}_+; f(x) < k+1\} \geq M - k - 1$ implies that $f(M - k - 1) < k + 1$. Therefore $F(M - k) = f(M - k) + M - k \geq M$ and $F(M - k - 1) < M$. This means that exactly $M - k$ terms of the form $F(n)$ are smaller than M; therefore $N = M$.

The converse is immediate by defining $f(n) = F(n) - n$.

Comments 41 *Since f^* is nondecreasing, it follows that it makes sense to define f^{**}. The sequences $G(n)$ and $H(n)$, where $H(n) = f^{**}(n) + n$, are complementary, and thus $f^{**} = f$.*

The result is due to A. Frankel.

Problem 2.29. Let M denote the set of bijective functions $f : \mathbb{Z}_+ \to \mathbb{Z}_+$. Prove that there is no bijective function between M and $\mathbb{Z}$.

Solution 2.29. We will find an injection $\rho : \mathbb{R} \setminus (\mathbb{Q} \cup [0, 2]) \to M$, as follows. To any $\alpha \in \mathbb{R} \setminus (\mathbb{Q} \cup [0, 2])$ there is associated an irrational $\beta < 2$ such that $\frac{1}{\alpha} + \frac{1}{\beta} = 1$. Let $\rho(\alpha) : \mathbb{Z}_+ \to \mathbb{Z}_+$ be the function defined by

$$\rho(\alpha)(n) = \begin{cases} \left[\alpha \frac{n}{2} \right], & \text{for even } n, \\ \left[\beta \frac{n+1}{2} \right], & \text{for odd } n. \end{cases}$$

According to the previous problem, $\rho(\alpha) : \mathbb{Z}_+ \to \mathbb{Z}_+$ is a bijection. Moreover, the map ρ is easily seen to be injective. Thus $\mathrm{card}(M) \geq \mathrm{card}(\mathbb{R} \setminus \mathbb{Q}) = \mathrm{card}(\mathbb{R}) > \mathrm{card}(\mathbb{Z})$, and the result follows.

Problem 2.30. Let $F \subset \mathbb{Z}$ be a finite set of integers satisfying the following properties:

1. For any $x \in F$, there exist $y, z \in F$ such that $x = y + z$.
2. There exists n such that, for any natural number $1 \leq k \leq n$, and any choice of $x_1, \ldots, x_k \in F$, their sum $x_1 + \cdots + x_k$ is nonzero.

Prove that $\mathrm{card}(F) \geq 2n + 2$.

Solution 2.30. By hypothesis, $0 \notin F$. Let $F_+ = F \cap \mathbb{Z}_+$, $F_- = F \cap \mathbb{Z}_-$, so that $F = F_+ \cup F_-$. Consider the unoriented graph Γ whose vertices are the elements of F_+ and whose edges are defined below: x and y are adjacent if there exists $z \in F$ such that $x = y + z$. By the first hypothesis, each vertex is adjacent to at least one edge. This implies that the graph Γ contains a cycle $[x_1, \ldots, x_k]$. Assume that k is minimal with this property. This means that there exist $z_i \in F$ such that:

$$x_1 = x_2 + z_1 = x_3 + z_2 + z_1 = \cdots = x_1 + z_k + \cdots + z_1,$$

which implies that

$$z_1 + \cdots + z_k = 0.$$

The second hypothesis implies that $k \geq n + 1$, and thus $\mathrm{card}(F_+) \geq k \geq n + 1$. A similar argument shows that $\mathrm{card}(F_-) \geq n + 1$, and the claim follows.

Problem 2.31. For a finite graph G we denote by $Z(G)$ the minimal number of colors needed to color all its vertices such that adjacent vertices have different colors. This is also called the chromatic number of G.

Prove that the inequality

$$Z(G) \geq \frac{p^2}{p^2 - 2q}$$

holds if G has p vertices and q edges.

Solution 2.31. If we fix $N = \chi(G)$ and the number p of vertices of the graph G, then we have to show that the number q of edges allowed

$$q \leq \frac{p^2}{2}\left(1 - \frac{1}{N}\right).$$

By hypothesis, there exists a partition (according to the colors) of the set of vertices into N classes such that two vertices in the same class are not adjacent. Let $n_1, n_2, \ldots, n_N$ be the number of vertices in the respective classes. The only possibility to get an edge is to join two vertices lying in different classes, and thus the total number q of edges is at most $\sum_{1 \leq i < j \leq N} n_i n_j$. Further, one knows that $\sum_{1 \leq i \leq N} n_i = p$. We now have

$$\sum_{i<j} n_i n_j \leq C_N^2 \frac{p^2}{N^2} = \frac{p^2}{2}\left(1 - \frac{1}{N}\right).$$

The equality holds if and only if N divides p and $n_i = p/N$, for all i.

Comments 42 *It can be proved, in the same way, that we have*

$$\chi(G) \geq \frac{p}{[t]}\left(1 - \frac{t - [t]}{1 + [t]}\right),$$

where $t = p - \frac{2q}{p}$. There exists also an upper bound for an arbitrary graph, namely

$$\chi(G) \leq \frac{1}{2}\left(1 + \sqrt{8q + 1}\right).$$

One problem that received a great deal of attention in the past was to get bounds for the chromatic number of a graph in terms of its topology. For instance, assume that the graph G is drawn on some surface so that its edges are not crossing each other, in which case the graph is said to be embedded. If the surface is partitioned into curvilinear polygons (called "countries"), then the dual graph of this decomposition is constructed by associating a vertex to each country and joining two vertices if the respective countries have a common frontier. This way one constructs all graphs embedded in that surface. Surfaces in $\mathbb{R}^3$ are characterized by only one number, called the genus g, or equivalently, by its Euler–Poincaré characteristic, which is $2 - 2g$. The latter is computed elementarily starting from an arbitrary partition into curvilinear polygons. If such a partition consists of V vertices, E edges, and F polygonal countries, then

$$2 - 2g = V - E + F.$$

Computing a bound for the chromatic number of an arbitrary graph embedded in a surface is equivalent to finding the number of colors needed for coloring any map of the surface so that adjacent countries have different colors. Now, if G is a graph that can be embedded in a surface of genus g, then it was stated without proof by Heawood in 1890 that

$$\chi(G) \le \left[\frac{1}{2}(7 + \sqrt{1 + 48g})\right].$$

This has been known as the Heawood conjecture since then, and it was finally proved in 1968 for all surfaces except for the plane by Ringel and Youngs. The case of the plane remained open until 1977, when Appel and Haken presented a computer-assisted proof of that case, thenceforth known as the "four color theorem." The article of Stahl presents the history of the problem and of the proofs.

- K. Appel, W. Haken: *Every Planar Map is Four Colorable,* With the collaboration of J. Koch, Contemporary Mathematics, 98, American Mathematical Society, Providence, RI, 1989.
- G. Ringel: *Map Color Theorem,* Die Grundlehren der mathematischen Wissenschaften, Band 209, Springer-Verlag, 1974.
- S. Stahl: *The other map coloring theorem,* Math. Magazine 58 (1985), 3, 131–145.

Problem 2.32. Let D_k be a collection of subsets of the set $\{1, \ldots, n\}$ with the property that whenever $A \ne B \in D_k$, then card$(A \cap B) \le k$, where $0 \le k \le n - 1$. Prove that

$$\text{card}(D_k) \le C_n^0 + C_n^1 + C_n^2 + \cdots + C_n^{k+1}.$$

Solution 2.32. Let P_k be the set of those subsets of $\{1, \ldots, n\}$ with at most k elements. Then any maximal (with respect to inclusion) collection D_k as above must contain P_k, since otherwise, $D_k \cup P_k$ will still satisfy the requirements while being strictly larger than D_k. Therefore, let us consider a maximal D_k and set $\widetilde{D}_k = D_k \setminus P_k$. For $A \in \widetilde{D}_k$, let us set $\widetilde{D} = \{A_1, \ldots, A_s \mid \text{card}(A_i) = k + 1, \text{ and } A_i \subset A\}$. If $B \in \widetilde{D}_k$ and $\widetilde{A} \cap \widetilde{B} \ne \emptyset$, then there exist i, j such that $A_i = B_j$, where card$(A_i) = k + 1$; this would imply that $A \cap B \supset A_i$ and thus card$(A \cap B) \ge k + 1$, which is absurd.

Therefore, the sets $\widetilde{A}$, for $A \in \widetilde{D}_k$, are disjoint. Since each $\widetilde{A}$ has at least one element and $\widetilde{A} \subset P_{k+1} \setminus P_k$, we obtain that card$(\widetilde{D}_k) \le$ card $(P_{k+1} \setminus P_k)$ and hence card$(D_k) \le$ card$(\widetilde{D}_k) +$ card$(P_k) =$ card$(P_{k+1}) = C_n^0 + C_n^1 + C_n^2 + \cdots + C_n^{k+1}$. We have equality for $D_k = P_{k+1}$.

Problem 2.33. Prove that

$$\frac{1}{p!}\sum_{k=0}^{n}(-1)^{n-k}C_n^k k^p = \begin{cases} 0, & \text{if } 0 \le p < n, \\ 1, & \text{if } p = n, \\ n/2, & \text{if } p = n + 1, \\ \frac{n(3n+1)}{24}, & \text{if } p = n + 2, \\ \frac{n^2(n+1)}{48}, & \text{if } p = n + 3, \\ \frac{n(15n^3+30n^2+5n+1)}{1152}, & \text{if } p = n + 4. \end{cases}$$

Solution 2.33. Consider the function $(e^x - 1)^n$, in which we develop first the binomial and then we develop the exponentials in Taylor series:

$$(e^x - 1)^n = \sum_{k=0}^{n} (-1)^{n-k} C_n^k e^{kx} = \sum_{k=0}^{n} (-1)^{n-k} C_n^k \left(\sum_{j=0}^{\infty} \frac{1}{j!} k^j x^j \right)$$

$$= \sum_{j=0}^{\infty} \frac{1}{j!} \left(\sum_{k=0}^{n} (-1)^{n-k} C_n^k k^j \right).$$

For the same function we now develop the exponential and then the binomial as follows:

$$(e^x - 1)^n = \left(x + \frac{1}{2!} x^2 + \frac{1}{3!} x^3 + \cdots \right)^n$$

$$= x^n + \frac{n}{2} x^{n+1} + \frac{n(3n+1)}{24} x^{n+2} + \frac{n^2(n+1)}{48} x^{n+3}$$

$$+ \frac{n(15n^3 + 30n^2 + 5n + 1)}{1152} x^{n+4} + \cdots.$$

By identifying the first coefficients in the two series, we obtain the formulas from the statement.

Comments 43 *The problem of computing the sums $s_{n,p} = \sum_{k=0}^{n} (-1)^k C_n^k k^p$ is a classical one, and the first result (for $p = n$) is due to Leonhard Euler. D. Andrica gave a method of computation by means of the recurrence relations*

$$s_{n,p+1} = n(s_{n,p} - s_{n-1,p}), \quad p \geq 0.$$

In particular, one recovers the results from the problem.

- D. Andrica: *On a combinatorial sum*, Gazeta Mat. 5 (1989), 158.

Problem 2.34. Write $\mathrm{lcm}(a_1, \ldots, a_n)$ in terms of the various $\gcd(a_i, \ldots, a_j)$ for subsets of $\{a_1, \ldots, a_n\}$.

Solution 2.34. Let $P_j = \prod_{1 \leq i_1 < \cdots < i_j \leq n} \gcd(a_{i_1}, \ldots, a_{i_j})$. We have

$$\mathrm{lcm}(a_1, \ldots, a_n) = \frac{P_1 P_3 P_5 \cdots}{P_2 P_4 P_6 \cdots}.$$

For instance, if $n = 3$ we have

$$\mathrm{lcm}(x, y, z) = \frac{xyz \gcd(x, y, z)}{\gcd(x, y)\gcd(y, z)\gcd(z, x)}.$$

Let p be a prime and e_i maximal such that p^{e_i} divide a_i. We can assume that $e_n \leq e_{n-1} \leq \ldots \leq e_1$. In P_j, the prime p appears to the e_ith power in C_{n-1}^{j-1} cases, and so the exponent of p in $\prod_i P_{2i+1} / \prod_i P_{2i}$ is e_1, because the alternating sum of the binomial coefficients is zero, with the exception of $C_0^0 = 1$.

Problem 2.35. Let $f(n)$ be the number of ways in which a convex polygon with $n+1$ sides can be divided into regions delimited by several diagonals that do not intersect (except possibly at their endpoints). We consider as distinct the dissection in which we first cut the diagonal a and next the diagonal b from the dissection in which we first cut the diagonal b and next the diagonal a. It is easy to compute the first values of $f(n)$, as follows: $f(1) = 1$, $f(2) = 1$, $f(3) = 3$, $f(4) = 11$, $f(5) = 45$. Find the generating function $F(x) = \sum f(n)x^n$ and an asymptotic formula for $f(n)$.

Solution 2.35. By a simple counting argument, we obtain the recurrence

$$f(n) = 3f(n-1) + 2\sum_{k=2}^{n-2} f(k)f(n-k).$$

If we write $y = F(x) - x$, then y satisfies the functional equation $y = x^2 + 3xy + 2y^2$. This is a consequence of the recurrence relation above. Solving the degree-two equation in y, we derive that

$$F(x) = \frac{1}{4}\left(1 + x - \sqrt{1 - 6x + x^2}\right).$$

Using Taylor's formula and induction on n, one further shows that

$$f(n) = \frac{\sqrt{3\sqrt{2} - 4}}{4\sqrt{\pi}} \cdot \frac{\left(3 + 2\sqrt{2}\right)^n}{n\sqrt{n}}\left(1 + \frac{3(8 - 3\sqrt{2})}{32n} + \mathcal{O}\left(n^{-2}\right)\right).$$

Problem 2.36. Find the permutation $\sigma : (1, \dots, n) \to (1, \dots, n)$ such that

$$S(\sigma) = \sum_{i=1}^{n} |\sigma(i) - i|$$

is maximal.

Solution 2.36. We want to prove that a permutation σ that maximizes $S(\sigma)$ satisfies $\sigma(1) > \sigma(2) > \cdots > \sigma(n)$.

Let us prove first the following intermediate result. If $a < c, b < d$, then the inequality

$$|a - b| + |c - d| \le |a - d| + |b - c|$$

holds with equality when either $a < c < b < d$ or $b < d < a < c$.

From the symmetry of the previous inequality, we can assume that $a < b, d$, and using a translation followed by a homothety on the real axis, we can, moreover, assume that $a = 0, c = 1$.

If $b < 1, d < 1$, the inequality reduces to $2b < 2d$.

If $b < 1, d > 1$, then the inequality is $2b < 2$.

If $b > 1, d > 1$, we obtain equality. This proves the claim.

Now let σ be a permutation such that $S(\sigma)$ is maximal. Suppose that there exists $i < j$ such that $\sigma(i) < \sigma(j)$. Let us define another permutation $\sigma^* = \sigma \circ (i, j)$, where (i, j) denotes the transposition interchanging i and j.

According to our claim, we have $S(\sigma^*) \geq S(\sigma)$.

By modifying iteratively σ by taking the product with the transpositions determined by all pairs (i, j) as above, we will finally obtain γ:

$$\gamma : \begin{pmatrix} 1 & 2 & \cdots & n \\ n & n-1 & \cdots & 1 \end{pmatrix}.$$

Moreover, the inequality above shows that

$$S(\sigma) \leq S(\gamma) = \sum_{i=1}^{n} |n - 2i + 1|.$$

Problem 2.37. On the set S_n of permutations of $\{1, \ldots, n\}$ we define an invariant distance function by means of the formula

$$d(\sigma, \tau) = \sum_{i=1}^{n} |\sigma(i) - \tau(i)|.$$

What are the values that d could possibly take?

Solution 2.37. We have $d(\rho\sigma, \rho\tau) = d(\sigma, \tau)$, for any $\rho, \sigma, \tau \in S_n$. Therefore, it suffices to consider the values of $d(\mathbf{1}, \sigma)$, where $\mathbf{1}$ is the identity permutation.

If $m > 0$ is a value of d, we will show below that $m - 2$ is also a value of d, and hence d takes all even values from 0 to some $2t$, where t has to be determined.

Notice first that d takes only even values, because

$$d(\sigma, \tau) \equiv \sum_{i=1}^{n} (\sigma(i) - \tau(i)) \equiv \sum_{i=1}^{n} \sigma(i) - \sum_{i=1}^{n} \tau(i) \equiv 0 \pmod{2}.$$

Let $m = d(\mathbf{1}, \sigma) > 0$. There then exist $1 \leq r < s \leq n$ such that $\sigma(r) > r, \sigma(s) < s$, and $\sigma(i) = i$, if $r < i < s$.

Let $\rho_{r,s}$ be the cycle $(r, r+1, \ldots, s)$. We claim that $d(\mathbf{1}, \rho_{r,s}\sigma) = m - 2$. First, by hypothesis, $\sigma(r) > s$ and $\sigma(s) < r$. Looking at the contribution of the elements from r to s to the respective distances, we obtain

$$d(\mathbf{1}, \sigma) - d(\mathbf{1}, \rho_{r,s}\sigma) = (\sigma(r) - r + s - \sigma(s)) - (r - \sigma(s) + s - r - 1 + \sigma(r) - r - 1) = 2.$$

Let us show now that $t = \left[\frac{n^2}{4}\right]$ (see also the previous problem) and the maximum distance $d(\mathbf{1}, \sigma)$ is realized for the permutation $\theta(i) = n + 1 - i$.

We have $d(\mathbf{1}, \theta) = 2\left[\frac{n^2}{4}\right] = \left[\frac{n^2}{2}\right]$. Let $k < n - 1$, such that $\sigma(i) = n + 1 - i$, for all $i < k$, while $\sigma(k) \neq n + 1 - k$. We claim that there exists some permutation $\tilde{\sigma} \in S_n$ with the property that $d(\mathbf{1}, \sigma) \leq d(\mathbf{1}, \tilde{\sigma})$ and that satisfies $\tilde{\sigma}(i) = n + 1 - i, i \leq k$. Then, using induction k, we find a sequence of permutations reaching θ such that

$$d(\mathbf{1}, \sigma) \le d(\mathbf{1}, \tilde{\sigma}) \le \cdots \le d(\mathbf{1}, \theta).$$

If $\sigma(k) \ne n + 1 - k$, we have $\sigma(k) < n + 1 - k$ because every number greater than $n + 1 - k$ is the image of some $i < k$. But this implies that $\sigma(r) = n + 1 - k$, for some $r > k$. Let τ be the transposition (k, r). Set $\tilde{\sigma} = \tau\sigma$; we have

$$d(\mathbf{1}, \tilde{\sigma}) - d(\mathbf{1}, \sigma) = |r - k\sigma| + |m + 1 - 2k| - |k - k\sigma| + |n + 1 - k - r| = w.$$

Let us suppose that $k \le \sigma(k) \le r \le n + 1 - k$ (the other cases are similar). Then $w = r - \sigma(k) + n + 1 - 2k - \sigma(k) + k - n - 1 + k + r = 2(r - k\sigma) \ge 0$. Analogously, for all k, r we have

$$d(\mathbf{1}, \tilde{\sigma}) - d(\mathbf{1}, \sigma) \ge 0,$$

which proves the claim.

Problem 2.38. The set $M = \{1, 2, \ldots, 2n\}$ is partitioned into k sets $M_1, \ldots, M_k$, where $n \ge k^3 + k$. Show that there exist $i, j \in \{1, \ldots, k\}$ for which we can find $k + 1$ distinct even numbers $2r_1, \ldots, 2r_{k+1} \in M_i$ with the property that $2r_1 - 1, \ldots, 2r_{k+1} - 1 \in M_j$.

Solution 2.38. There exists a set M_s that contains at least $\frac{2n}{k} \ge 2(k^2 + 1)$ elements. We have to consider two cases:

1. Either M_s contains at least $\frac{2(k^2+1)}{2} = k^2 + 1$ even numbers. Then the set of odd numbers

$$O = \{r - 1, \text{ where } r \text{ is even, and } r \in M_s\}$$

has $k^2 + 1$ elements. Then there exists some set M_a containing at least $\frac{k^2+1}{k}$ elements from O. We choose therefore $i = s$ and $j = a$. Notice that i might be equal to j.

2. Or else M_s contains at least $k^2 + 1$ odd numbers. The solution is similar to that from above, considering the set of even numbers.

Problem 2.39. Let S be the set of odd integers not divisible by 5 and smaller than $30m$, where $m \in \mathbb{Z}_+^*$. Find the smallest k such that every subset $A \subset S$ of k elements contains two distinct integers, one of which divides the other.

Solution 2.39. Consider the subset $N = \{1, 7, 11, 13, 17, 19, 23, 29, \ldots, 30m - 1\}$ of elements of S that are not divisible by 3. There are $8m$ elements in N, which are written in increasing order $a_1 < a_2 < \cdots < a_{8m}$. Every element of S can be uniquely written as $x = a_i \cdot 3^t$, where $t \in \mathbb{Z}_+$ and $a_i \in N$.

If $k \ge 8m + 1$, then according to the pigeonhole principle, any subset $A \subset S$ of cardinality card $A = k$ contains two distinct elements $x, y \in A$ for which $x = a_i 3^t$ and $y = a_i 3^q$. In this case, either x divides y, or y divides x.

Next, for all i, choose the maximal $t(i)$ with the property that $a_i 3^{t(i)} < 30m < a_i 3^{t(i)+1}$, and set $b_i = a_i 3^{t(i)}$. We have therefore $10m < b_i < 30m$. The set $\{b_1, b_2, \ldots, b_{8m}\}$ contains $8m$ elements from S, and also we have the inequalities $0 < b_i/b_j < 3$, for any i, j. Since all numbers b_i are odd, we derive that $b_i/b_j \notin \mathbb{Z}$. Therefore $\{b_1, \ldots, b_{8m}\}$ does not contain a number and one divisor of it. This shows that the required value of k is $8m + 1$.

Problem 2.40. Prove that $\prod_{1\le j<i\le n} \frac{a_i-a_j}{i-j}$ is a natural number whenever $a_1 \le a_2 \le \cdots \le a_n$ are integers.

Solution 2.40. We can suppose $a_i \ge 0$. Then,

$$S = \prod_{i>j} \frac{a_i - a_j}{i - j} = \frac{\prod_{i>j}(a_i - a_j)}{(n-1)!(n-2)!\cdots 1!}.$$

We now consider the determinant D of the matrix

$$\begin{pmatrix} 1 & 1 & \cdots & 1 \\ C_{a_1}^1 & C_{a_2}^1 & \cdots & C_{a_n}^1 \\ \vdots & \vdots & & \vdots \\ C_{a_1}^{n-1} & C_{a_2}^{n-1} & \cdots & C_{a_n}^{n-1} \end{pmatrix}.$$

It is obvious that $D \in \mathbb{Z}_+$. We now write the binomial coefficient C_m^k using factorials, and we arrange the factors in the lines. We obtain

$$D = \frac{1}{1!2!\cdots (n-1)!} \det \begin{pmatrix} 1 & 1 & \cdots & 1 \\ a_1 & a_2 & \cdots & a_n \\ \vdots & \vdots & & \vdots \\ a_1^{n-1} & a_2^{n-1} & \cdots & a_n^{n-1} \end{pmatrix} = \frac{\prod_{i>j} a_i - a_j}{\prod_{i=1}^n (i-1)!} = S,$$

and the claim follows.

Problem 2.41. Is there an infinite set $A \subset \mathbb{Z}_+$ such that for all $x, y \in A$ neither x nor $x + y$ is a perfect power, i.e., a^k, for $k \ge 2$? More generally, is there an infinite set $A \subset \mathbb{Z}_+$ such that for any nonempty finite collection $x_i \in A, i \in J$, the sum $\sum_{i\in J} x_i$ is not a perfect power?

Solution 2.41. 1. Yes. Let $p_1 = 2, p_2 = 3, \ldots, p_n, \ldots$ be the set of primes in increasing order. We then set $A = \{2^2 3, 2^2 3^2 5, \ldots, 2^2 3^2 5^2 \cdots p_n^2 p_{n+1}, \ldots\}$.

Obviously, no element of A is a perfect power. Moreover, if $x \le y \in A$, then we can write $x = 2^2 \cdots p_k^2 p_{k+1}$ and $x = 2^2 \cdots p_n^2 p_{n+1}$. If $k < n$, then p_{k+1} divides $x + y = 2^2 \cdots p_k^2 p_{k+1}(1 + p_{k+1} p_{k+2}^2 \cdots p_n^2 p_{n+1})$, while p_{k+1}^2 does not divide $x + y$, and thus it cannot be a perfect power. If $k = n$, the same argument works, unless $k = 0$, which was excluded from A.

2. Let us prove that if $C \subset \mathbb{Z}_+$ is a set of density $d(C) = 0$, then there exists an infinite set $A \subset \mathbb{Z}_+$ such that for any nonempty finite collection $x_i \in A, i \in J$, the sum $\sum_{i\in J} x_i$ is not in C.

Recall that the density of the subset $C \subset \mathbb{Z}_+$ is defined as

$$d(C) = \lim_{n\to\infty} \frac{\text{card}(C \cap \{1, 2, \ldots, n\})}{n}.$$

Since $d(C) = 0$, there exists a natural number $a_1 \notin C$. Then consider $C_1 = C \cup (C - a_1) \cap \mathbb{Z}_+$, where we set $X - \lambda = \{x - \lambda | x \in X\}$. The density d has the following fundamental property that results from the definition:

$$d(A \cup B) \leq d(A) + d(B).$$

This implies that $d(C_1) = 0$, and therefore there exists a natural number $a_2 \notin C_1$.

Assuming that both the subset C_k of density zero and the sequence $a_1, a_2, \ldots, a_{k+1}$, with $a_{k+1} \notin C_k$, are defined, we set

$$C_{k+1} = C_k \bigcup_{p=1}^{k} \bigcup_{i_1 \leq i_2 \leq \cdots \leq i_p \leq k+1} (C_k - (a_{i_1} + a_{i_2} + \cdots + a_{i_p})) \cap \mathbb{Z}_+.$$

It follows that $d(C_{k+1}) = 0$ and thus there exists at least one integer $a_{k+2} > a_{k+1}$ such that $a_{k+2} \notin C_{k+1}$.

The sequence (a_k) forms an infinite subset A with the required property.

It suffices now to compute the density of the set $C = \{a^k | a \geq 2, k \geq 2\}$. We have

$$\frac{\text{card}(C \cap \{1, 2, \ldots, n\})}{n} = \frac{[\sqrt{n}] + [\sqrt[3]{n}] + \cdots}{n} \leq \frac{\sqrt{n} \log_2 n}{n} = \frac{\log_2 n}{\sqrt{n}}.$$

Therefore $d(C) = 0$.

Comments 44 *There is an analogous result when products are used instead of sums of elements in A. For instance, the set*

$$A = \left\{ a_1, a_2 = C_{2a_1}^{a_1}, C_{a_1+a_2}^{a_2}, \ldots, C_{a_{n-1}+a_n}^{a_n} \right\}$$

has the property that no product of its elements is a perfect power. Is it true that $d(A) = 0$?

Problem 2.42. Let

$$f(n) = \max A_1^{A_2^{A_3^{\cdot^{\cdot^{\cdot^{A_k}}}}}},$$

where $n = A_1 + \cdots + A_k$. Thus, $f(1) = 1$, $f(2) = 2$, $f(3) = 3$, $f(4) = 4$, $f(5) = 9$, $f(6) = 27$, $f(7) = 512$, etc. Determine $f(n)$.

Solution 2.42. It is clear that if $f(n) = A_1^{A_2^{\cdot^{\cdot^{\cdot^{A_k}}}}}$, then $A_i > 1$. Also, f is an increasing function, and thus $f(n+1) \geq f(n) + 1$. If $A_1 = k$, then $A_2^{\cdot^{\cdot^{\cdot^{A_k}}}} \leq f(n-k)$, and this implies that whenever $n \geq 4$, we have

$$f(n) = \max_{2 \leq k \leq n-2} k^{f(n-k)}.$$

It is easy now to see that $f(n+1) \geq 2f(n)$. In fact, let $f(n) = k^{f(n-k)}$, for some $k \geq 2$; then

$$f(n+1) \geq k^{f(n-k+1)} \geq k^{f(n-k)+1} \geq 2f(n).$$

The next step is to compare $a^{f(b+1)}$ with $(a+1)^{f(b)}$, that is, $\frac{f(b+1)}{f(b)}$ with $\frac{\log(a+1)}{\log(a)}$. If $a \geq 2$, then $a + 1 < a^2$ and hence $\log(a+1)/\log a < 2$. Thus, if $b \geq 4$, then

$f(b + 1)/f(b) \geq 2$, which yields $a^{f(b+1)} > (a + 1)^{f(b)}$, as soon as $a \geq 2, b \geq 3$. For small values of the arguments, we have

$$\frac{\log 4}{\log 3} < \frac{f(4)}{f(3)} < \frac{f(3)}{f(2)} < \frac{\log 3}{\log 2}.$$

These show that $k^{f(n-k)}$ has a maximum when $k = 2$ if $n > 6$. Therefore, the final answer is

$$f(n) = \begin{cases} 2^{\cdot^{\cdot^{2^{3^2}}}}, & \text{for odd } n > 3, \\ 2^{\cdot^{\cdot^{2^{3^3}}}}, & \text{for even } n > 4. \end{cases}$$

Problem 2.43. Consider a set M with m elements and $A_1, \ldots, A_n$ distinct subsets of M such that $\text{card}(A_i \cap A_j) = r \geq 1$ for all $1 \leq i \neq j \leq n$. Prove that $n \leq m$.

Solution 2.43. Observe first that if A is an $m \times n$ matrix and $m > n$, then $\det(A \cdot A^\top) = 0$, where $A^\top$ denotes the transpose matrix. In fact, let us border the matrix A by a null matrix of size $(m - n) \times n$, in order to get an $m \times m$ matrix A'. It is obvious that $A' \cdot A'^\top = A \cdot A^\top$, and thus $\det(A \cdot A^\top) = \det(A' \cdot A'^\top) = 0$.

Let D be the $n \times n$ matrix whose entries are

$$D_{ij} = \begin{cases} r \geq 1, & \text{if } i \neq j, \\ d_{ii} \geq r, & \text{otherwise.} \end{cases}$$

Assume that there exists at most one i such that $d_{ii} = r$. We claim then that $\det(D) \neq 0$. This follows by subtracting the last line from the first $n - 1$ and an inductive argument.

Consider now the $m \times n$ matrix A whose entries are

$$A_{ij} = \begin{cases} 1, & \text{if } i \in A_j, \\ 0, & \text{otherwise.} \end{cases}$$

It is now immediate that $D = A \cdot A^\top$ is given by

$$D_{ij} = \sum_{j=1}^n a_{ik} a_{kj} = \text{card}(A_i \cap A_j).$$

By hypothesis, $D_{ij} = r$ when $i \neq j$, and since the A_i are distinct, there exists at most one i such that $D_{ii} = r$.

The claim from above tells us that $\det(D) = \det(A \cdot A^\top) \neq 0$, while the first observation implies that $m \leq n$.

Problem 2.44. Set $\pi(n)$ for the number of prime numbers less than or equal to n. Prove that there are at most $\pi(n)$ numbers $1 < a_1 < \cdots < a_k \leq n$ with $\gcd(a_i, a_j) = 1$.

Solution 2.44. If X is a set, we denote by $\mathcal{P}(X)$ the set of nonempty subsets of X. Define the map $h : \{1, \ldots, n\} \to \mathcal{P}(\{p_1, \ldots, p_{\pi(n)}\})$ that associates to the integer x the set of prime divisors of x.

One then has $\mathrm{card}(h(a_i)) \geq 1$ and further $h(a_i) \cap h(a_j) = \emptyset$, from our assumption. Since the maximal number of disjoint subsets of $\{p_1, \ldots, p_{\pi(n)}\}$ is bounded by $\pi(n)$, the claim follows.

Problem 2.45. Prove that for every k, there exists n such that the nth term of the Fibonacci sequence F_n is divisible by k. Recall that F_n is determined by the recurrence $F_{n+2} = F_n + F_{n+1}$, for $n \geq 0$, where the first terms are $F_0 = 0$, $F_1 = 1$.

Solution 2.45. Using the pigeonhole principle, there exists an infinite sequence n_i such that

$$F_{n_1} \equiv F_{n_2} \equiv \cdots \equiv F_{n_m} \equiv \cdots \pmod{k}.$$

Moreover, there exists an infinite subsequence n_{i_s} such that

$$F_{n_{i_1}+1} \equiv F_{n_{i_2}+1} \equiv \cdots \equiv F_{n_{i_m}+1} \equiv \cdots \pmod{k}$$

by the same argument. Thus one knows that

$$F_{n_{i_s}} \equiv F_{n_{i_t}} \pmod{k} \text{ and } F_{n_{i_s}+1} \equiv F_{n_{i_t}+1} \pmod{k}.$$

The recurrence relation of the Fibonacci numbers shows that

$$F_{n_{i_s}+m} \equiv F_{n_{i_t}+m} \pmod{k}, \text{ for any } m \in \mathbb{Z}.$$

Since $F_0 = 0$, we find that there exists n (actually infinitely many such integers) such that k divides F_n.

Problem 2.46. Consider a set of consecutive integers $C + 1, C + 2, \ldots, C + n$, where $C > n^{n-1}$. Show that there exist distinct prime numbers $p_1, p_2, \ldots, p_n$ such that $C + j$ is divisible by p_j.

Solution 2.46. Let $k \leq h$ and consider the factorization of $C + k$ into prime factors. If $C + k$ has at least n distinct prime divisors, then we can find one prime divisor that is not yet associated with the other $n - 1$ numbers $C + i$. Assume then that $C + k$ has at most $j \leq n - 1$ prime factors, which are $p_{1k}, \ldots, p_{jk}$ for $1 \leq j \leq n - 1$. We write down the factorization as

$$p_{1k}^{a_{1k}} \cdots p_{jk}^{a_{jk}} = C + k > C > n^{n-1}.$$

The pigeonhole principle implies that there exists some prime power $q_k = p_{jk}^{a_{jk}}$ such that $q_k > n$.

We then associate the prime number p_{jk} to $C + k$. Let us show that this procedure yields distinct prime numbers. Suppose that there exist j and k such that the associated numbers are the same. We know that q_k divides $C + k$ and q_j divides $C + j$, and they are both prime powers of the same prime. Since $q_k > n$ and $q_j > n$, we obtain $\gcd(q_k, q_j) > n$. Moreover, $\gcd(q_k, q_j)$ divides both $C + k$ and $C + j$ and hence their difference $|j - k| < n$, which is a contradiction. This proves the claim.

Problem 2.47. Let p be a prime number and $f(p)$ the smallest integer for which there exists a partition of the set $\{2, 3, \ldots, p\}$ into $f(p)$ classes such that whenever $a_1, \ldots, a_k$ belong to the same class of the partition, the equation

$$\sum_{i=1}^{k} x_i a_i = p$$

does not have solutions in nonnegative integers. Estimate $f(p)$.

Solution 2.47. 1. Suppose that a_1, a_2 are in the same class and $\gcd(a_1, a_2) = 1$. One knows then that any natural number greater than or equal to $a_1 a_2$ can be written as $x_1 a_1 + x_2 a_2$, where $x_i \in \mathbb{Z}_+$. Therefore, the prime numbers smaller than $\sqrt{p}$ should to different classes of our partition, and hence

$$f(p) > \pi\left(p^{\frac{1}{2}}\right) \sim \frac{2\sqrt{p}}{\log p},$$

where $\pi(n)$ denotes the number of primes smaller than n.

2. If $t \in \mathbb{Z}_+$, let us consider the sets of consecutive elements $A_t = \left\{\left[\frac{p}{t+1}\right] + 1, \ldots, \left[\frac{p}{t}\right]\right\}$. We claim that the equation $\sum_{i=1}^{k} a_i x_i = p$ has no integer solutions if $a_i \in A_t$ for $1 \leq i \leq k$. In fact, if we had a solution x_i, then we would have

$$\frac{p}{t+1} \sum_{i=1}^{k} x_i < \sum_{i=1}^{k} a_i x_i < \frac{p}{t} \sum_{i=1}^{k} x_i,$$

and so $t < \sum_{i=1}^{k} x_i < t + 1$, which is false, since x_i are integers.

Let us consider the classes $A_1, A_2, \ldots, A_{L-1}$, where L is an integer to be fixed later.

Consider now, for every prime $q < p/L$, the classes $B_q = \{q, 2q, \ldots, \alpha q, \ldots\}$. It is immediate that the associated linear equation has no solutions if the coefficients belong to some B_q, since q does not divide p. Further, $A_1, \ldots, A_{L-1}, B_2, B_3, \ldots, B_{\left[\frac{p}{L}\right]}$ form a partition of $\{2, 3, \ldots, p\}$. The total number of classes of this partition is then $L - 1 + \pi(p/L)$.

Setting $L = \sqrt{\frac{2p}{\log p}}$, we obtain $f(p) < \sqrt{\frac{8p}{\log p}}(1 + \mathcal{O}(1))$.

Problem 2.48. Consider m distinct natural numbers a_i smaller than N such that $\text{lcm}(a_i, a_j) \leq N$ for all i, j. Prove that $m \leq 2\left[\sqrt{N}\right]$.

Solution 2.48. Assume that the numbers a_i are ordered as $1 \leq a_m < \cdots < a_1 \leq N$. We will prove by induction on k that $a_k \leq \frac{N}{k}$. If $k = 1$, then it is obvious. Moreover, if $k \geq 1$, then

$$a_k - a_{k+1} \geq \gcd(a_k, a_{k+1}) = \frac{a_k a_{k+1}}{\text{lcm}(a_k, a_{k+1})} \geq \frac{a_k a_{k+1}}{N}.$$

This is equivalent to $a_{k+1} + \frac{a_k a_{k+1}}{N} \leq a_k$, which yields

$$a_{k+1} \le \frac{N a_k}{a_k + N} = N - \frac{N^2}{a_k + N} \le N - \frac{N^2}{\frac{N}{k} + N} = \frac{N}{k+1},$$

and our claim follows. Furthermore, one has

$$a_{[\sqrt{N}]+1} \le \frac{N}{[\sqrt{N}]+1} \le [\sqrt{N}].$$

The sequence (a_j) contains then at most $[\sqrt{N}]$ terms between 1 and $[\sqrt{N}]$. On the other hand, between $\sqrt{N}$ and N, there are no more than $[\sqrt{N}]$ terms of our sequence, and thus we have a maximum number of $m \le 2[\sqrt{N}]$ terms.

Comments 45 *If $m(N)$ denotes the maximum number of terms of a sequence (a_i), as in the statement, then Erdős conjectured the following asymptotic behavior of the function $m(n)$:*

$$m(N) = \frac{3}{2^{3/2}} \sqrt{N} + \mathcal{O}(1).$$

See also:

1. P. Erdős: *Remarks on number theory. IV. Extremal problems in number theory,* Matematikai Lapok 13 (1962), 228–255.

Problem 2.49. The set $M \subset \mathbb{Z}_+$ is called A-sum-free, where $A = (a_1, a_2, \ldots, a_k) \in \mathbb{Z}_+^k$, if for any choice of $x_1, x_2, \ldots, x_k \in M$ we have $a_1 x_1 + a_2 x_2 + \cdots + a_k x_k \notin M$. If A, B are two vectors, we define $f(n; A, B)$ as the greatest number h such that there exists a partition of the set of consecutive integers $\{n, n+1, \ldots, h\}$ into S_1 and S_2 such that S_1 is A-sum-free and S_2 is B-sum-free. Assume that $B = (b_1, b_2, \ldots, b_m)$ and that the conditions below are satisfied:

$$a_1 + a_2 + \cdots + a_k = b_1 + b_2 + \cdots + b_m = s,$$

and

$$\min_{1 \le j \le k} a_j = \min_{1 \le j \le m} b_j = 1, \quad k, m \ge 2.$$

Prove that $f(n; A, B) = ns^2 + n(s-1) - 1$.

Solution 2.49. First consider the sets

$$S_1 = \{n, n+1, \ldots, ns-1\} \cup \{ns^2, ns^2+1, \ldots, ns^2 + n(s-1) - 1\},$$
$$S_2 = \{ns, ns+1, \ldots, ns^2 - 1\}.$$

If $x_j \in S_2$, then $b_1 x_1 + b_2 x_2 + \cdots + b_m x_m \ge ns^2$ and thus S_2 is B-sum-free.
   If $x_1, x_2, \ldots, x_k \in S_1$, then we have two cases:

1. If $x_i \le ns - 1$ for all i, then

$$ns \le a_1 x_1 + a_2 x_2 + \cdots + a_k x_k \le s(ns-1) < ns^2.$$

2. If some x_i belongs to $\{ns^2, ns^2 + 1, \ldots, h\}$, then

$$a_1 x_1 + a_2 x_2 + \cdots + a_k x_k \geq ns^2 + n(s-1).$$

Thus S_1 is A-sum-free. This proves that $f(n; A, B) \geq ns^2 + n(s-1) - 1$.

Assume now that the set $\{n, n+1, \ldots, ns^2 + n(s-1)\}$ can be partitioned into two sets S_i that are A (respectively B) sum-free.

Let $n \in S_1$. If we take $x_1 = x_2 = \cdots = x_k = n$, then $ns = a_1 x_1 + a_2 x_2 + \cdots + a_k x_k \notin S_1$, so $ns \in S_2$. Taking now $x_1 = x_2 = \cdots + = x_m = ns \in S_2$, we find that $ns^2 \in S_1$.

Suppose that $a_1 = b_1 = 1$. If we consider the elements $x_1 = ns^2, x_2 = \cdots = x_k = n$ of S_1, we derive that $n(s^2 + s - 1) \in S_2$.

1. If $n(s+1) \in S_1$, then take $x_1 = n$, $x_2 = \cdots = x_k = n(s+1)$, and since $ns^2 \in S_1$, we find that the set S_1 is not A-sum-free.
2. If $n(s+1) \in S_2$, then take $x_1 = ns$, $x_2 = \cdots = x_m = n(s+1)$, and since $n(s^2 + s - 1) \in S_2$, we obtain that the set S_2 is not B-sum-free.

This contradiction shows that $f(n; A, B) = ns^2 + n(s-1) - 1$.

Comments 46 *Generalized sum-free sets of integers were considered first by Rado, who in 1933 gave the upper bound*

$$f(1; A, A) \leq \max\left((bmc^2 - 1)(c-1) + bmc, \frac{bmc^2(c-1)}{a} \right),$$

where $A = (a, b)$, $m = \frac{a}{\gcd(a,b)}$ *and* $c = \max(x_0, y_0, z_0)$, *where* (x_0, y_0, z_0) *is the minimal solution of the Diophantine equation* $ax + by = z$. *There are several partial results known for particular forms of A and B. If we denote the vector $(d, d, \ldots, d)$ with k components by $k\langle d\rangle$, then Kasá proved that*

$$f(1; 2\langle 1\rangle, k\langle 1\rangle) = \begin{cases} 3k - 3, & \text{for odd } k, \\ 3k - 2, & \text{for even } k, \end{cases}$$

$$f(n; 2\langle 1\rangle, k\langle 1\rangle) = (2k+1)n - 1, \quad \text{for even } n.$$

Further, Seress improved this by showing that

$$f(n; m\langle 1\rangle, k\langle 1\rangle) = (mk + m - 1)n - 1, \quad \text{for } n > 1, m \geq 3.$$

The result from the problem is due to L. Funar, who proved also that

$$f(n; k\langle d\rangle, k\langle d\rangle) = k^2 d^3 + kd - d - 1$$

for d even and $k \geq d$ and in several other cases. Abbott showed that this formula holds for all $k, d \geq 2$. Other estimates for $f(n; A, B)$ in the case in which A, B have not necessarily the same sum of components have been obtained by P. Moree. These suggest that there is no such simple closed formula for arbitrary A, B. However, recent progress has lead to the determination of $f(1; A, A)$ for arbitrary A (also *known as the 2-color Rado number), by Guo and Sun (improving previous results by B. Hopkins and D. Schaal). Specifically, we have*

$$f(1; A, A) = a(s-a)^2 + (2a^2+1)(s-a) + a^3,$$

where $a = \min_{1\le j\le k} a_j$ and $s = a_1 + \cdots + a_k$.

The general case, when we consider partitions into $n \ge 3$ subsets S_j such that each S_j is A_j-sum-free, seems to be much more difficult, and no general estimates from above are known.

- H.L. Abbott: *On a conjecture of Funar concerning generalized sum-free sets,* Nieuw Arch. Wisk. (4) 9 (1991), 249–252.
- L. Funar: *Generalized sum-free sets of integers,* Nieuw Arch. Wisk. (4) 8 (1990), 49–54.
- Song Guo and Zhi-Wei Sun: *Determination of the two-color Rado number for $a_1x_1 + \cdots + a_mx_m = x_0$,* math.CO/0601409.
- B. Hopkins and D. Schaal: *On Rado numbers for $\sum_{i=1}^{m-1} a_i x_i = x_m$,* Adv. Appl. Math. 35 (2005), 433–441.
- P. Moree: *On a conjecture of Funar,* Nieuw Arch. Wisk. (4) 8 (1990), 55–60.
- R. Rado: *Studien zur Kombinatorik,* Math. Zeitschrift 36(1933), 424–480.
- A. Seress: *k-sum-free decompositions,* Matematikai Lapok 31 (1978/83), 191–194.

Problem 2.50. Let $1 \le a_1 < a_2 < \cdots < a_n < 2n$ be a sequence of natural numbers for $n \ge 6$. Prove that

$$\min_{i,j} \operatorname{lcm}(a_i, a_j) \le 6\left(\left[\frac{n}{2}\right]+1\right).$$

Moreover, the constant 6 is sharp.

Solution 2.50. We assume that among the a_i's we can find both a and $2a$. Then $\min_{i,j} \operatorname{lcm}(a_i,a_j) \le \operatorname{lcm}(a,2a) \le 2n < 6\left(\left[\frac{n}{2}\right]+1\right)$. Let $a \le n$ be an element of the sequence, if it exists. If $2a$ also belongs to the sequence, the claim follows from above. If $2a$ is not in the sequence, then we will replace the element a with $2a$. This way the value of $\min_{i,j} \operatorname{lcm}(a_i, a_j)$ is not diminished. We continue this process as far as possible. At some point, all the elements of the sequence will be greater than n, and thus the last sequence is forced to be $n+1, n+2, \ldots, 2n$.

If $n = 2k+1$, then take $a = 2k+2$, $b = 3k+3$, so that $\operatorname{lcm}(a,b) = 6k+6 = 6\left(\left[\frac{n}{2}\right]+1\right)$.

Now if c,d are such that $n+1 < c < d \le 2n$, then we have $\operatorname{lcm}(c,d) \ge 6\left(\left[\frac{n}{2}\right]+1\right)$. In fact, $\operatorname{lcm}(c,d) = pd = qc$ for some $q > p > 1$, and thus either $p = 2$ and $q \ge 3$, or else $q \ge 4$.

In the first case, c is an even number; hence $c \ge 2\left(\left[\frac{n}{2}\right]+1\right)$ and $\operatorname{lcm}(c,d) = qc \ge 3c \ge 6\left(\left[\frac{n}{2}\right]+1\right)$.

In the second case, $\operatorname{lcm}(c,d) = qc \ge 4(n+1) > 6\left(\left[\frac{n}{2}\right]+1\right)$. Therefore, if we choose the set $n+1, \ldots, 2n$, we have $\min_{i,j} \operatorname{lcm}(a_i, a_j) = 6\left(\left[\frac{n}{2}\right]+1\right)$.

Comments 47 *This result is due to P. Erdős, who claimed also that under the same hypothesis, we have*

$$\max_{i \neq j} \gcd(a_i, a_j) > \frac{38n}{147} - C,$$

where C is a constant independent of n.

Problem 2.51. Let $1 \leq a_1 < a_2 < \cdots < a_k < n$ be such that $\gcd(a_i, a_j) \neq 1$ for all $1 \leq i < j \leq k$. Determine the maximum value of k.

Solution 2.51. Let $f(n)$ denote the maximum of k as a function of n. We have then $f(2) = 1$. Assume from now on that $n \geq 3$.

Define l_i (for $1 \leq i \leq k$) to be the smallest integer satisfying the inequality $n/2 < a_i 2^{l_i}$. If there exists a pair of distinct indices i and j such that $a_i 2^{l_i} = a_j 2^{l_j}$, then either a_i divides a_j, or conversely, a_j divides a_i. Therefore, we can replace a_i by $a_i 2^{l_i}$ without diminishing the value of k.

Thus we can assume that $a_i \geq \frac{n}{2}$, for all i. Further, the condition $\gcd(a_i, a_j) \neq 1$ tells us that we cannot find two consecutive numbers that both belong to the sequence a_i. Thus, if n is of the form $4m - 1, 4m$, or $4m + 2$, then $f(n) \leq m$.

Moreover, if $n = 4m + 1$, then $f(n) \leq m + 1$. If we have equality above, then the sequence a_i is forced to be $a_1 = 2m + 1, a_2 = 2m + 3, \ldots, a_{m+1} = 4m + 1$, and therefore $\gcd(a_1, a_2) = 1$, which is a contradiction. Thus $f(n) \leq m$ if $n = 4m + 1$.

By considering the set of all even numbers in the interval $[n/2, n]$ we obtain that for any $n \geq 3$, we have $f(n) = \left\lceil \frac{n+1}{4} \right\rceil$.

Problem 2.52. Consider the increasing sequence $f(n) \in \mathbb{Z}_+, 0 < f(1) < f(2) < \cdots < f(n) < \cdots$. It is known that the nth element in increasing order among the positive integers that are not terms of this sequence is $f(f(n)) + 1$. Find the value of $f(240)$.

Solution 2.52. 1. One knows that the nth absent number is $N = f(f(n)) + 1$. However, the numbers smaller than N that belong to the sequence are $f(1), f(2), \ldots, f(f(n))$. Moreover, there are n numbers less than or equal to N that do not belong to the sequence. This means that $N = f(n) + n$; thus

$$f(f(n)) = f(n) + n - 1.$$

Therefore $f(1) = 1, f(2) = 3$. It follows by induction that

$$f(n) = n + \operatorname{card} \{m \mid f(m) < n\}.$$

Set $a_0 = 2, a_{n+1} = f(a_n)$. We compute

$$a_{n+1} = a_n + \operatorname{card} \{m \mid f(m) < a_n\} = a_n + \operatorname{card} \{m \mid f(m) < a_{n-1}\} = a_n + a_{n-1} - 1,$$

and therefore the sequence $b_n = a_n - 1$ satisfies the Fibonacci recurrence $b_{n+1} = b_n + b_{n-1}$. We will prove by induction that

$$f(b_n + x) = b_{n+1} + f(x), \text{ for } 1 \leq x \leq b_{n-1}.$$

If $n = 0$, then $b_1 = 1$ and $x = 1$ and the claim is verified. Further

$$f(b_n + x) = b_n + x + \text{card } \{y | f(y) < f(b_{n-1})\} + \text{card } \{y | b_{n+1} < f(y) < b_n + x\},$$

which implies that $f(b_n - 1) < f(a_{n-1}) = a_n$ and thus $f(b_{n-1}) \leq b_n \leq b_n + x$. In particular,

$$f(y) \leq b_n + x \leq b_n + b_{n-1} \leq b_{n+1} < f(a_n),$$

from which we obtain $y < a_n$, $y \leq b_n$. We now write

$$y = b_{n+1} + z, \text{ where } z \leq b_n - b_{n-1} = b_{n-2}.$$

Hence

$$f(b_n + x) = b_n + x + b_{n-1} + \text{card } \{z | f(z) < x\} = b_{n+1} + f(x).$$

This proves our claim.

Recall now that every natural number can be uniquely written as a sum of Fibonacci numbers, i.e., as

$$x = 1 + b_{k_1} + b_{k_2} + \cdots + b_{k_p},$$

where $1 \leq k_1 < k_2 < \cdots < k_p$. If x is written as above, the previous inductive formula yields the value

$$f(x) = 1 + b_{k_1+1} + b_{k_2+1} + \cdots + b_{k_p+1}.$$

In particular,

$$f(240) = 1 + 2 + 8 + 377 = 388,$$

since $240 = 1 + (1 + 5 + 233)$.

2. We can use in a clever manner the recursion law $f(f(n)) = f(n) + n - 1$ in order to find, step by step,

$$f(3) = 3 + 1 = 4, \quad f(4) = 4 + 2 = 6, \quad f(6) = 6 + 3 = 9,$$
$$f(9) = 9 + 5 = 14, \quad f(14) = 22$$
$$f(22) = 35, \quad f(35) = 56, \quad f(56) = 90, \quad f(90) = 145,$$
$$f(145) = 234, \quad f(234) = 378.$$

One knows that $f(f(35)) + 1 = 91$ is absent, so that $f(57) = 92$ and thus $f(92) = 148$, $f(148) = 239$, $f(239) = 386$. Next $f(f(148)) + 1 = 387$ is absent, and hence $f(240) = 388$.

Problem 2.53. We define inductively three sequences of integers (a_n), (b_n), (c_n) as follows:

1. $a_1 = 1$, $b_1 = 2$, $c_1 = 4$;
2. a_n is the smallest integer that does not belong to the set
   $\{a_1, \ldots, a_{n-1}, b_1, \ldots, b_{n-1}, c_1, \ldots, c_{n-1}\}$;

3. b_n is the smallest integer that does not belong to the set
$$\{a_1, \ldots, a_{n-1}, a_n, b_1, \ldots, b_{n-1}, c_1, \ldots, c_{n-1}\};$$
4. $c_n = 2b_n + n - a_n$.

Prove that
$$0 < n\left(1 + \sqrt{3}\right) - b_n < 2 \quad \text{for all } n \in \mathbb{Z}_+.$$

Solution 2.53. We will actually prove that

$$\alpha < n\left(1 + \sqrt{3}\right) - b_n < \beta, \quad \text{where } \alpha = (9 - 5\sqrt{3})/3, \ \beta = (12 - 4\sqrt{3})/3.$$

Observe that the sequence (c_n) does not contain two consecutive numbers. Thus a_{n+1} might jump at most two units ahead from b_n, and b_n two units ahead from a_n. By induction, we obtain

$$1 \leq b_n - a_n \leq 2 \text{ and } 1 \leq a_{n+1} - b_n \leq 2.$$

Using the equality

$$c_{n+1} - c_n = 2(b_{n+1} - a_{n+1}) + (a_{n+1} - b_n) - (b_n - a_n) + 1$$

we deduce that
$$2 \leq c_{n+1} - c_n \leq 6.$$

In particular, there are not six consecutive integers among the sequences a_j and b_j.

Set $\gamma_n = n(1 + \sqrt{3}) - b_n$. We will prove by induction that $\alpha < \gamma_n < \beta$, which is trivially verified for $n = 1$. Suppose now that this holds true for all $n < k$. We consider the truncated sequence $a_{k-2} < b_{k-2} < a_{k-1} < b_{k-1} < a_k < b_k$. As remarked above, there should be at least one element of the form c_j, which can be inserted between a_{k-2} and b_k, since $b_k - a_{k-2} \geq 6$. Let us choose the greatest such c_n, which is $c_n \leq b_k - 1$.

It follows that the set

$$\{a_1, a_2, \ldots, a_k, b_1, b_2, \ldots, b_k, c_1, c_2, \ldots, c_n\}$$

is precisely the set of the first consecutive numbers contained between 1 and b_k. In particular, we have

$$b_k = 2k + n = c_n + r, \quad \text{where } 1 \leq r \leq 5.$$

On the other hand, we know that $c_n = 2b_n + n - a_n = b_n + n + (b_n - a_n)$ and $b_n - a_n \leq 2$, and so

$$2k + n = b_k = b_n + n + s, \quad \text{where } 2 \leq s \leq 7.$$

However, one can improve the upper bound for s as follows. We have equality only if $r = 5$, but in this case we have $c_{n+1} - c_n = 6$. Using the formula for $c_{n+1} - c_n$, we derive that $b_n - a_n = 1$, and thus one has $s \leq 6$.

Let us now estimate γ_k, which reads

$$\gamma_k = (1+\sqrt{3})k - b_k = (1+\sqrt{3})\left(\frac{b_n + s}{2}\right) - (b_n + n + s) = (s - \gamma_n)\left(\frac{\sqrt{3}-1}{2}\right).$$

If $s \leq 6$ we use $\gamma_n \geq \alpha$ to obtain

$$\gamma_k \leq (6 - \alpha)\left(\frac{\sqrt{3}-1}{2}\right) < \beta.$$

If $s = 6$, then we must have $b_{n+1} - a_{n+1} = a_{n+1} - b_n = 2$, and therefore $b_{n+1} - b_n = 4$. Thus $\gamma_{n+1} - \gamma_n = 1 + \sqrt{3} - (b_{n+1} - b_n) = \sqrt{3} - 3$. Since $n + 1 < k$, one uses the induction hypothesis $\gamma_{n+1} > \alpha$ in order to get $\gamma_n = \gamma_{n+1} + 3 - \sqrt{3} > \alpha + 3 - \sqrt{3}$. Using this inequality above, we derive that

$$\gamma_k \leq (6 - \alpha - 3 + \sqrt{3})\left(\frac{\sqrt{3}-1}{2}\right) = \beta.$$

If $s = 5$, then $b_{n+1} - b_n \in \{3, 4\}$. Thus $\gamma_{n+1} - \gamma_n \leq \sqrt{3} - 2$, and so $\gamma_n > \alpha + 2 - \sqrt{3}$. Introducing this above, we obtain

$$\gamma_k < (5 - \alpha - 2 + \sqrt{3})\left(\frac{\sqrt{3}-1}{2}\right) = \beta.$$

The other inequality, $\gamma_k > \alpha$, follows along the same lines.

6.3 Geometric Combinatorics

Problem 2.54. We consider n points in the plane that determine C_n^2 segments, and to each segment one associates either $+1$ or -1. A triangle whose vertices are among these points will be called negative if the product of numbers associated to its sides is negative. Show that if n is even, then the number of negative triangles is even. Moreover, for odd n, the number of negative triangles has the same parity as the number p of segments labeled -1.

Solution 2.54. Let P_i denote the points in the plane, and let a_{ij} be the label of the segment $P_i P_j$. Then the signature of the triangle $P_i P_j P_k$ is given by $P_{ijk} = a_{ij} a_{jk} a_{ki}$. In particular,

$$\prod_{i \neq j \neq k \neq i} P_{ijk} = (-1)^m,$$

where m is the number of negative triangles. Now every segment belongs to $n - 2$ triangles, and thus

$$\prod_{i \neq j \neq k \neq i} P_{ijk} = \prod a_{ij} a_{jk} a_{ki} = \left(\prod_{i \neq j} a_{ij}\right)^{n-2} = (-1)^{p(n-2)}.$$

Therefore $m \equiv pn \pmod 2$, and the claim follows.

Problem 2.55. Given n, find a finite set S consisting of natural numbers larger than n, with the property that, for any $k \geq n$, the $k \times k$ square can be tiled by a family of $s_i \times s_i$ squares, where $s_i \in S$.

Solution 2.55. Each of the sets $S = \{s \in \mathbb{Z} | \ n \leq s \leq n^2\}$, and $S = \{s \in \mathbb{Z} | s$ is prime, $n^2 < s < 2n^2 + n\}$ is convenient, although neither one is minimal.

We will prove this by induction on the size of the square, for the second set S. Consider a $k \times k$ square, where $k \geq n^2$. By hypothesis, the $m \times m$ square is tiled by squares from S, for any m with $n \leq m < k$. If k is composite, then write $k = pq$, where $k > p \geq n$. We cover the $k \times k$ square by means of $q^2 \ p \times p$ squares and we proceed inductively. If k is prime, then $k > 2n^2 + n$. We divide the square into two squares, one $m \times m$ and the other $(k - m) \times (k - m)$, and two $m \times (k - m)$ rectangles, where $m = n(n + 1)$. We have $k - m > n^2$ and thus each $m \times (k - m)$ rectangle can be covered with $m \times n$ or $m \times (n + 1)$ pieces and each of these pieces can be further divided into squares.

Comments 48 *The problem whether some region can be tiled by a given set of tiles is difficult and unsolved in general. The interested reader might consult the survey of Ardila and Stanley.*

The simpler problem of whether an $n \times m$ rectangle was tiled by $a \times b$ rectangles was solved by de Bruijn and, independently, by Klarner in 1969. Specifically, this happens to be possible if and only if mn is divisible by ab, both m and n can be written as sums of a's and b's, and either m or n is divisible by a or else m (or n) is divisible by b. More generally, the $m_1 \times m_2 \times \cdots \times m_n$ box is tiled by $a_1 \times a_2 \times \cdots \times a_n$ bricks if and only if each of the a_i has a multiple among the m_j's. Conway and Lagarias gave necessary conditions for the existence of tilings of rectilinear polygons (i.e., those having sides parallel to the axes) by a set of finite rectilinear tiles using boundary invariants, which are combinatorial group-theoretic invariants associated with the boundaries of the tile shapes and the regions to be tiled. Some new invariants were discovered recently by Pak.

Notice that for $n \geq 2, m > 2$, the problem whether there exists a tiling of a given rectilinear polygon having only horizontal $n \times 1$ tiles and only vertical $1 \times m$ tiles is an NP-complete question, as shown by Beauquier et al.

- F. Ardila, R. Stanley: *Tilings,* math.CO/0501170.
- D. Beauquier, M. Nivat, E. Rémila, M. Robson: *Tiling figures of the plane with two bars,* Comput. Geom. 5 (1995), 1–25.
- N. de Bruijn: *Filling boxes with bricks,* Amer. Math. Monthly 79 (1969), 37–40.
- J.H. Conway, J.C. Lagarias: *Tiling with polyominoes and combinatorial group theory,* J. Combin. Theory Ser. A 53 (1990), 183–208.
- D. Klarner: *Packing a rectangle with congruent n-ominoes,* J. Combin. Theory 7 (1969), 107–115.
- I. Pak: *Ribbon tile invariants,* Trans. Amer. Math. Soc. 352 (2000), 12, 5525–5561.

Problem 2.56. We consider $3n$ points $A_1, \ldots, A_{3n}$ in the plane whose positions are defined recursively by means of the following rule: first, the triangle $A_1 A_2 A_3$ is equilateral; further, the points A_{3k+1}, A_{3k+2}, and A_{3k+3} are the midpoints of the sides of the triangle $A_{3k}A_{3k-1}A_{3k-2}$. Let us assume that the $3n$ points are colored with two colors. Show that for $n \geq 7$ there exists at least one isosceles trapezoid having vertices of the same color.

Solution 2.56. Any subset of three points has two points of the same color. Therefore, there are two points having the same color (which we denote by A_{i_k} and A_{j_k}) among $\{A_{3k+1}, A_{3k+2}, A_{3k+3}\}$.

Consider now the lines determined by these monocolor pairs of points:

$$\{A_{i_1}A_{j_1}, A_{i_2}A_{j_2}, \dots, A_{i_n}A_{j_n}\}.$$

We know that all these directions are parallel to the three sides of the initial equilateral triangle. Thus there exist at least $[\frac{n}{3}] + 1 \geq 3$ lines in the set above that are parallel. Moreover, each line comes with one of the two colors, and thus there exist at least $\left[\frac{[\frac{n}{3}]+1}{2} \right] + 1 \geq 2$ parallel lines of the same color. The four points that determine these lines form a monocolor isosceles trapezoid.

Problem 2.57. Is there a coloring of all lattice points in the plane using only two colors such that there are no rectangles with all vertices of the same color, whose side ratio belongs to $\left\{1, \frac{1}{2}, \frac{1}{3}, \frac{2}{3}\right\}$?

Solution 2.57. No. According to Van der Waerden's theorem, there exist four lattice (equidistant) points $A_i = (x_i, 0)$ of the same color (say red) on the axis $y = 0$ such that $x_2 - x_1 = x_3 - x_2 = x_4 - x_3 = k$.

Consider the points B_i with coordinates (x_i, k), lying on a parallel line, at distance k. Then the points B_j contain either two red points—in which case we are done, by considering the respective A_j—or else at least three black points, B_{j_1}, B_{j_2}, and B_{j_3}.

Consider next the points $C_i = (x_i, 2k)$. Then there exist two points among C_{j_1}, C_{j_2}, and C_{j_3} that have the same color.

If this color is black, then using the respective B_{j_i} and C_{j_i}, we obtain a black rectangle as needed. If the color is red, then the respective A_{j_i} and C_{j_i} form a red rectangle, as claimed.

Comments 49 *Van der Waerden's theorem from 1927 states that if the set of natural numbers $\mathbb{Z}_+$ is partitioned into finitely many color classes, then there exist monochromatic arithmetic progressions of arbitrary length. Erdős and Turán conjectured in 1936 the existence of arbitrarily long arithmetic progressions in sequences of integers with positive density, and their conjecture was successfully confirmed by Szemerédi in 1975. Many extensions have been made since that time. Actually, Szemerédi proved that for a given length l and density $\delta > 0$, there is an $L(l, \delta)$ such that if $L \geq L(l, \delta)$, any subset $A \subset \{1, 2, 3, \dots, L\}$ with more than δL elements, will contain an arithmetic progression of length l. This makes precise Van der Waerden's theorem in that there exists $W(l, r)$ such that if $W \geq W(l, r)$ and $\{1, 2, 3, \dots, W\}$ is the union of r subsets, then one of these necessarily contains an l-term arithmetic progression, and moreover, $W(l, r)$ can be taken as $L(l, 1/r)$.*

The next breakthrough was made recently by Gowers, who provided the first effective upper bound to the function $L(l, \delta)$. One consequence is a significant improvement of the upper bound for $W(l, r)$; for example, for some constant c, $W(4, r)$ can be taken as $\exp(\exp(r^c))$. Generally, we can take $L(l, \delta) = \exp(\delta^{-c(l)})$, where $c(l) = 2^{2^{l+9}}$. These estimates fall short of providing a proof to another conjecture of Erdős claiming that any set $A \subset \mathbb{N}$ with $\sum_{a \in A} \frac{1}{a} = \infty$ contains arbitrarily long arithmetic progressions.

Another far-reaching generalization was provided by Bergelson and Leibman, as follows. Suppose that $p_1, p_2, \ldots, p_m$ are polynomials with integer coefficients and no constant term. Then, whenever $\mathbb{Z}_+$ is finitely colored, there exist natural numbers a and d such that the point a and all the points $a + p_i(d)$, for $1 \le i \le m$, have the same color.

Finally, Szemerédi's theorem was proved to hold for the primes in a spectacular recent paper by Ben Green and Terence Tao. Specifically, if A is a subset of positive density within the set of primes (for instance the set of all primes), then A contains infinitely many arithmetic progressions of length k (for any $k \ge 3$). An exposition of this result can be found in the survey of Kra.

- V. Bergelson, A. Leibman: *Polynomial extensions of Van der Waerden's and Szemerédi's theorems*, J. Amer. Math. Soc. 9 (1996), 725–753.
- W.T. Gowers: *A new proof of Szemerédi's theorem*, Geometric Functional Analysis 11 (2001), 465–588.
- B. Kra: *The Green–Tao theorem on arithmetic progressions in the primes: an ergodic point of view*, Bull. Amer. Math. Soc. (N.S.) 43 (2006), 1, 3–23.
- E. Szemerédi: *On sets of integers containing no k elements in arithmetic progression*, Collection of articles in memory of J.V. Linnik, Acta Arith. 27 (1975), 199–245.
- B.L. Van der Waerden: *Beweis einer Baudetschen Vermutung*, Nieuw Arch. v. Wiskunde 15 (1927), 212–216.

Problem 2.58. Let G be a planar graph and let P be a path in G. We say that P has a (transversal) self-intersection in the vertex v if the path has a (transversal) self-intersection from the curve-theoretic viewpoint. Let us give an example. Take the point 0 in the plane and the segments 01, 02, 03, 04 going counterclockwise around 0. Then a path traversing first 103 and then 204 has a (transversal) self-intersection at 0, while a path going first along 102 and further on 304 does not have a (transversal) self-intersection.

Prove that any connected planar graph G, with only even-degree vertices, admits an Eulerian circuit without self-intersections. Recall that an Eulerian circuit is a path along the edges of the graph, that passes precisely once along each edge of the graph.

Solution 2.58. We will proceed by double induction, first on the maximum degree of the vertices and, for the class of graphs with several vertices of maximum degree, on the number of such vertices. If the degree is $d = 2$, then the obvious Eulerian circuit does not have self-intersections. Now let v be a vertex of maximal degree $d > 2$.

We split v into two vertices, one of degree 2 and the other of degree $d - 2$; the first among the new vertices is incident to two vertices that were previously incident to v, by means of two consecutive edges (this makes sense since the graph is planar), and the latter is incident to the remaining ones. If the obtained graph is connected, then there exists an Eulerian path without self-intersections, and then we simply restore the graph and see that this way the circuit remains without self-intersections. If the graph is not connected, then we get two Eulerian circuits. When restoring the graph G, we glue together the two circuits into one Eulerian circuit, still without self-intersections.

Problem 2.59. Let us consider finitely many points in the plane that are not all collinear. Assume that one associates to each point a number from the set $\{-1, 0, 1\}$ such that the following property holds: for any line determined by two points from the set, the sum of numbers associated to all points lying on that line equals zero. Show that, if the number of points is at least three, then to each point one associates the number 0.

Solution 2.59. Let us assume the contrary, i.e., that there exist points with nontrivial numbers associated to them. Then there should exist points labeled both 1 and -1.

Choose first a point x_0 labeled 1. For any other point $x \neq x_0$, the line $d = x_0 x$ has $n_\varepsilon(d)$ points labeled by $\varepsilon \in \{-1, 1\}$. Moreover, the assumptions imply that $n_{-1}(d) = n_{+1}(d)$ and $n_{+1}(d) \geq 1$, because x_0 is labeled 1.

Furthermore, the total number of negative points is $n_{-1} = \sum_d n_{-1}(d)$, the sum being taken on all lines d passing through x_0. Next, the total number of positive points is $n_{+1} = 1 + \sum_d (n_{+1}(d) - 1) = n_{-1} - l + 1$, where l is the number of distinct lines passing through x_0. We know that $l \geq 2$ by hypothesis; hence we obtain that $n_{+1} \leq n_{-1} - 1$.

Choose now a point y_0 labeled -1. Then the same argument implies that the reversed inequality $n_{-1} \leq n_{+1} - 1$ holds, contradicting the former inequality.

Problem 2.60. If one has a set of squares with total area smaller than 1, then one can arrange them inside a square of side length $\sqrt{2}$, without any overlaps.

Solution 2.60. We put the squares in decreasing order with respect to their side lengths. The first square is put in the left corner of the square S (the side of which is $\sqrt{2}$), the second square will be set to the right of S, and we go on until it becomes impossible to put another square without surpassing the borders of S. Then, we start another line of squares above the first square and so on. Assume that the side lengths are $S_1 \geq S_2 \geq \cdots$ and we have $n_1, n_2, \ldots, n_k$ squares on the lines number $1, 2, \ldots, k$ respectively. It follows that

$$\sqrt{2} - S_{n_1+1} \leq S_1 + S_2 + \cdots + S_{n_1} \leq \sqrt{2},$$
$$\sqrt{2} - S_{n_1+n_2+1} \leq S_{n_1+1} + \cdots + S_{n_1+n_2} \leq \sqrt{2},$$

and so on. We want to show that

$$S_1 + S_{n_1+1} + S_{n_1+n_2+1} + \cdots \leq \sqrt{2},$$

which will prove our claim. We know from above that

$$S_2 + S_3 + \cdots + S_{n_1+1} \geq \sqrt{2} - S_1,$$

from which we infer

$$S_2^2 + S_3^2 + \cdots + S_{n_1+1}^2 \geq \left(\sqrt{2} - S_1\right) S_{n_1+1},$$

and by the same trick, we obtain

$$S_{n_1+2}^2 + \cdots + S_{n_1+n_2+n_3}^2 \geq \left(\sqrt{2} - S_1\right) S_{n_1+n_2+1},$$

and so on. By summing up these inequalities, we obtain

$$1 - S_1^2 \geq S_2^2 + S_3^2 + \cdots \geq \left(\sqrt{2} - S_1\right)(S_{n_1+1} + S_{n_1+n_2+1} + \cdots),$$

and hence

$$S_1 + S_{n_1+1} + S_{n_1+n_2+1} + \cdots \leq \frac{1 - S_1^2}{\sqrt{2} - S_1} + S_1 = \sqrt{2} - \frac{\left(1 - \sqrt{2}S_1\right)^2}{\sqrt{2} - S_1} \leq \sqrt{2}.$$

Comments 50 *The problem of packing economically unequal rectangles into a given rectangle has recently received a lot of consideration. L. Moser asked in 1968 to find the smallest packing default ε in each of the following situations:*

1. *the set of rectangles of dimensions $1 \times \frac{1}{n}$, for $n \in \mathbb{Z}_+, n \geq 2$, whose total area equals 1 can be packed into the square of area $1 + \varepsilon$ without overlap;*
2. *the set of squares of side lengths $\frac{1}{n}$, for $n \in \mathbb{Z}_+$, can be packed without overlap into a rectangle of area $\pi^2/6 - 1 + \varepsilon$;*
3. *the (infinite) set of squares with sides of lengths $1/(2n + 1)$, $n \in \mathbb{Z}_+$, can be packed in a rectangle of area $\pi^2/8 - 1 + \varepsilon$.*

The best results known to this day yield ε very small, at least smaller than 10^{-9}, using a packing algorithm due to M. Paulhus. Further computational investigations support the apparently weaker claim that the packings in the problems above are possible for every positive number ϵ. However, G. Martin showed that if this weaker claim holds, then one can also find a packing for $\varepsilon = 0$, in which case the packing is called perfect. The general case of the problem above is still open, but a variation of it has been solved in the meantime. J. Wästlund proved that if $1/2 < t < 2/3$, then the squares of side n^{-t}, for $n \in \mathbb{Z}_+$, can be packed into some finite collection of square boxes of the same area $\zeta(2t)$ as the total area of the tiles. On the other hand, Chalcraft proved that there is a perfect packing of the squares of side $n^{-3/5}$ into a square, and presumably, his technique works for packing the squares of side n^{-t} into a square, where $1/2 < t \leq 3/5$; this is true for packing into a rectangle for all t in the range $0.5964 \leq t \leq 0.6$.

- A. Chalcraft: *Perfect square packings*, J. Combin. Theory Ser. A 92 (2000), 158–172.

- H.T. Croft, K.J. Falconer, R.K. Guy: *Unsolved Problems in Geometry,* Springer-Verlag, New York, 1994.
- G. Martin: *Compactness theorems for geometric packings,* J. Combin. Theory Ser. A 97 (2002), 225–238.
- A. Meir, L. Moser: *On packing of squares and cubes,* J. Combin. Theory 5 (1968), 126–134.
- M. Paulhus: *An algorithm for packing squares,* J. Combin. Theory Ser. A 82 (1998), 147–157.
- J. Wästlund: *Perfect packings of squares using the stack-pack strategy,* Discrete Comput. Geom. 29 (2003), 625–631.

Problem 2.61. Prove that for each k there exist k points in the plane, no three collinear and having integral distances from each other. If we have an infinite set of points with integral distances from each other, then all points are collinear.

Solution 2.61. 1. Consider the point P_1 on the unit circle having complex coordinates $z = \frac{3}{5} + \frac{4}{5}\sqrt{-1}$. If θ denotes the argument of P_1, then θ is noncommensurable with π, i.e., $\theta/\pi \in \mathbb{R} - \mathbb{Q}$. Consider the points P_n of coordinates z^n. Then the P_n form a dense set of points in the unit circle. Moreover, we can compute the distance

$$|P_n P_k| = 2|\sin(n - k)\theta|.$$

Since $\sin n\theta$ is a polynomial with rational coefficients in $\sin \theta$ and $\cos \theta$, we derive that the distances between all these points are rational. Given k, choose $P_1, \ldots, P_k$ and a homothety of large ratio in order to clear the denominators in these distances. We obtain then k points with integral pairwise distances.

2. Given three noncollinear points P, Q, and R, with $|PQ|, |PR| \leq k$, the number of points X such that the distances $|XP|, |XQ|, |XR|$ are integral is bounded by $4(k + 1)^2$. In fact, we have, by the triangle inequality,

$$\big||XP| - |XQ|\big| < |PQ| = k,$$

and thus

$$\big||XP| - |XQ|\big| \in \{1, 2, \ldots, k\}.$$

Moreover, the geometric locus of those X for which $\big||XP| - |XQ|\big| = j$ is a hyperbola with foci P and Q. On the other hand,

$$\big||XP| - |XR|\big| \in \{1, 2, \ldots, k\},$$

and thus X belongs to one of the $k + 1$ hyperbolas having P and R as foci. Thus X belongs to the intersections of $(k + 1)^2$ pairs of hyperbolas, which have at most $4(k + 1)^2$ intersection points.

Comments 51 *This construction of a dense set of points in the unit circle whose pairwise distances are rational is due to A. Muller. Then W. Sierpiński found such a dense set in the circle of radius r, provided that r^2 is rational, this condition being necessary as soon as we have three points at rational distances. The second result is due to Erdős. The question of Ulam whether there exists a dense set in the plane such that the distances between any two of its points is rational is still unsolved.*

- P. Erdős: *Integral distances,* Bull. Amer. Math. Soc. 51 (1945), 996.
- W. Sierpiński: *Sur les ensembles de points aux distances rationnelles situés sur un cercle,* Elem. Math. 14 (1959), 25–27.

Problem 2.62. Let O, A be distinct points in the plane. For each point x in the plane, we write $\alpha(x) = \widehat{xOA}$ (counterclockwise). Let $C(x)$ be the circle of center O and radius $|Ox| + \frac{\alpha(x)}{|Ox|}$. If the points in the plane are colored with finitely many colors, then there exists a point y with $\alpha(y) > 0$ such that the color of y also belongs to the circle $C(y)$.

Solution 2.62. Let G be the graph of vertices $\{x \in E^2, \alpha(x) > 0\}$ in which points x and y are adjacent if $y \in C(x)$.

If $C_1(\rho_1)$ and $C_2(\rho_2)$ are two circles of radii $\rho_1 < \rho_2 < 1$ centered at the origin, then there exists a point $M_1 \in C_1(\rho_1)$ that is adjacent to *all* points of $C_2(\rho_2)$. The point M_1 has $\alpha(M_1) = \theta$ uniquely determined, as follows. The condition $C(M_1) = C_2(\rho_2)$ is equivalent to $\rho_2 = \rho_1 + \frac{\theta}{\rho_1}$, yielding

$$\theta = (\rho_2 - \rho_1)\rho_1 \in (0, 1) \subset (0, 2\pi),$$

and therefore M_1 is well determined.

Assume now that the claim from the statement is not true. It will follow that the color of M_1 is distinct from the colors appearing on the circle $C_2(\rho_2)$.

Let now $C_1, C_2, \ldots, C_k, \ldots$ be an infinite sequence of circles centered at the origin whose respective radii are

$$0 < \rho_1 < \rho_2 < \cdots < \rho_k < \cdots < 1.$$

Then let the set of colors that we encounter on C_k be denoted by τ_k. The previous argument, when applied to the pair of circles C_i and C_j, shows that there exists a point on C_i that is colored by a color that does not exist on C_j, if $i < j$, and thus $\tau_i \setminus \tau_j \neq \emptyset$. Since the number of colors is finite, this is impossible as soon as k is large enough (e.g., $k > 2^p$, where p is the number of colors). This contradiction proves the claim.

Problem 2.63. Let $k, n \in \mathbb{Z}_+$.

1. Assume that $n-1 \leq k \leq \frac{n(n-1)}{2}$. Show that there exist n distinct points $x_1, \ldots, x_n$ on a line, that determine exactly k distinct distances $|x_i - x_j|$.
2. Suppose that $\left\lceil \frac{n}{2} \right\rceil \leq k \leq \frac{n(n-1)}{2}$. Then there exist n points in the plane that determine exactly k distinct distances.
3. Prove that for any $\varepsilon > 0$, there exists some constant $n_0 = n_0(\varepsilon)$ such that for any $n > n_0$ and $\varepsilon n < k < \frac{n(n-1)}{2}$, there exist n points in the plane that determine exactly k distinct distances.

Solution 2.63. 1. Let m be an integer between 1 and $n - 1$ such that

$$n - 1 = C_n^2 - C_{n-1}^2 \leq k \leq C_n^2 - C_m^2.$$

Set $p = k - (m - 1) - C_n^2 + C_{m+1}^2$; in particular, we have $1 \le p \le m$. Consider then the following set of points of the real line:

$$X = \{1, 2, \ldots, m, m + p\} \cup \{\pi^{m+2}, \pi^{m+3}, \ldots, \pi^n\}.$$

The distances between the first $m+1$ points correspond to the set $\{1, 2, \ldots, p+m-1\}$. The remaining points are independent transcendental points, and thus the distances between them are all distinct transcendental numbers. Moreover, the distances between points of the second set and points of the first set are also all distinct (translates of the former set of transcendental distances by integers). Thus the number of distances obtained so far is

$$p + m - 1 + \left(\sum_{i=m+2}^{n} (i - 1) \right) = k.$$

2. If $\left[\frac{n}{2}\right] \le k \le C_n^2$, we put the n points in the consecutive vertices of a regular $(2k + 1)$-gon. There are exactly k distinct distances in a regular $(2k + 1)$-gon, and all of them are realized at least once, since $k \ge \left[\frac{n}{2}\right]$.

3. It suffices to check the case $\varepsilon n \le k \le \left[\frac{n}{2}\right]$. Let $m = 3 + \left[\sqrt{n}\right]$. For $m \le a \le n - 2m$, $1 \le b \le m$, we consider the following subset of $\mathbb{Z} \times \mathbb{Z}$:

$$X = C \cup \{(i, 1)|1 \le i \le a - 1\} \cup \{(1, j)|2 \le j < m\} \cup \{(0, h)|1 \le h \le b\},$$

where C is a subset of $\{(1, j), 2 \le i \le a - 1, 2 \le j \le m\}$ with the property that the distances between the pairs of points in C are all distinct.

The set of squares of the distances between points in X is

$$\mathcal{D}(a, b) = \{(i^2 + j^2)|0 \le i \le a - 1, 0 \le j \le m - 1\} \cup \{a^2 + j^2|0 \le j \le b - 1\}.$$

We have

$$\mathcal{D}(m, 1) \subset \mathcal{D}(m, 2) \subset \cdots \subset \mathcal{D}(m, m) \subset \mathcal{D}(m + 1, 1)$$
$$\subset \mathcal{D}(m + 1, 2) \subset \cdots \subset \mathcal{D}(n - 2m, m),$$

and each set from above is obtained from the previous one in the sequence by adjoining at most one more distance; thus the difference between consecutive sets is either empty or a singleton. Therefore, card$(\mathcal{D}(a, b))$ is an increasing sequence of consecutive elements that lie between card$(\mathcal{D}(m, 1))$ and card$(\mathcal{D}(n - 2m, m))$. Observe now that the upper bound card$(\mathcal{D}(n - 2m, m))$ is at least $n - 2m - 1 \ge \left[\frac{n}{2}\right]$.

Further, the set $\mathcal{D}(m, 1)$ consists of integers between 1 and $2m^2$ (note that we have the asymptotic behavior $m^2 \sim n$) that are sums of two perfect squares. It is known that whenever $\mu = \lambda^2 + \delta^2$ and p is a prime number $p \equiv 3 \pmod 4$ if p divides μ, then p^2 also divides μ. Thus, the density of the set of numbers that are sums of two perfect squares is bounded by the product over the primes p

$$\prod_{p \equiv 3 \ (\mathrm{mod}\ 4)} \left(1 - \frac{1}{p} + \frac{1}{p^2} \right),$$

which tends to 0. Therefore, for n large enough, we have card$(\mathcal{D}(m, 1)) < \varepsilon n$, which proves the claim.

Comments 52 *The result is due to P. Erdős, who proposed it in 1980 as a problem in Amer. Math. Monthly, related to another question that he considered long ago. In 1946, Erdős posed the problem of determining the minimum number $d(n)$ of different distances determined by a set of n points in $\mathbb{R}^2$, proved that $d(n) \geq cn^{1/2}$, and conjectured that $d(n) \geq cn/\sqrt{\log n}$. If true, this inequality is best possible, as shown by the lattice points of the plane. The best result known is due to J. Solymosi and Cs.D. Tóth, and yields $d(n) > cn^{6/7}$.*

- P. Erdős: *On the set of distances of n points*, Amer. Math. Monthly 53 (1946), 248–250.
- P. Erdős: *Problem 6323*, Amer. Math.Monthly 87 (1980), 826.
- J. Solymosi, Cs.D. Tóth: *Distinct distances in the plane*, The Micha Sharir birthday issue, Discrete Comput. Geom. 25 (2001), 629–634.

Problem 2.64. Show that it is possible to pack $2n(2n + 1)$ nonoverlapping pieces having the form of a parallelepiped of dimensions $1 \times 2 \times (n + 1)$ in a cubic box of side $2n + 1$ if and only if n is even or $n = 1$.

Solution 2.64. Let us consider the cube composed of $(2n + 1)^3$ cells of dimensions $1 \times 1 \times 1$, which are labeled using the coordinate system $(a_1, a_2, a_3), a_i \in \{1, \ldots, 2n + 1\}$.

1. Case $n = 1$. The 6 pieces are arranged as follows. The first three pieces are given by

$$x_1 = \{(1, 1, 2), (1, 1, 3), (1, 2, 2), (1, 2, 3)\},$$
$$x_2 = \{(1, 2, 1), (1, 3, 1), (2, 2, 1), (2, 3, 1)\},$$
$$x_3 = \{(2, 1, 1), (3, 1, 1), (2, 1, 2), (3, 1, 2)\}.$$

The next three pieces x_4, x_5, x_6 are located as above, by permuting the indices 3 and 1. The cells $(1, 1, 1), (2, 2, 2), (3, 3, 3)$ remain empty.

2. Suppose now that n is even. We have then $(2n+1)^3 - 2n(2n+1) \cdot 1 \cdot 2 \cdot (n+1) = 2n + 1$ cases that remain empty. Each slice $P_{1,i} = \{(i, a, b) | a, b \in \{1, 2, \ldots, 2n+1\}\}$ can be divided in four smaller boxes $P_{1,i;1} = \{(i, a, b) | a \leq n + 1, b \leq n\}, P_{1,i;2} = \{(i, a, b) | n + 2 \leq a, b \leq n + 1\}, P_{1,i;3} = \{(i, a, b) | n + 1 \leq a, n + 2 \leq b\}, P_{1,i;4} = \{(i, a, b) | a \leq n, n + 1 \leq b\}$ and one extra cell, namely $\{(i, n + 1, n + 1)\}$.

Each box $P_{1,i;s}$ has dimensions $1 \times n \times (n + 1)$ and can be covered by $\frac{n}{2}$ pieces $1 \times 2 \times (n + 1)$. The remaining cells of coordinates $(i, n + 1, n + 1)$ remain empty.

3. Assume further that n is odd, $n \geq 3$, and that there exists a packing as in the statement. One associates to a cell (a_1, a_2, a_3) the number p, which counts the number of $i \in \{0, 1, 2, 3\}$ such that $a_i = n + 1$. One colors the cell blue if $p = 3$, red if $p = 2$, yellow when $p = 1$, and leaves it colorless when $p = 0$. Then, we have one blue cell and $6n$ red cells.

Now, for each $i \leq 2n + 1, k \in \{1, 2, 3\}$, we set $P_{k,i} = \{(a_1, a_2, a_3), a_k = i\}$.

Then each slice $P_{k,i}$ contains an odd number of cells $(2n + 1)^2$. In the meantime, every piece contains an even number of cells from a given slice $P_{k,i}$, which might be

$2, n + 1$, or $2(n + 1)$ cells, respectively. This means that in each slice $P_{k,i}$ there exists at least one empty cell.

The total number of empty cells is $2n+1$, and for fixed k the slices $P_{k,1}, \ldots, P_{k,2n+1}$ are disjoint, so that each slice $P_{k,i}$ contains exactly one empty cell. In particular, the set consisting of the blue cell and the red ones contains at most one empty cell.

Furthermore, each piece S contains colored cells because $n + 1 > \frac{1}{2}(2n + 1)$, and it fits into one of the cases below:

Type 1. S contains the blue cell, $n + 1$ red cells, and n yellow cells.

Type 2. S does not contain the blue cell, and it contains 2 red and $2n$ yellow cells.

Type 3. S does not contain the blue cell, and it contains 1 red, $n + 1$ yellow, and n colorless cells.

Type 4. S does not contain either the blue cell or red cells, but 2 yellow and $2n$ colorless cells.

Let us analyze now the total number of colored cells which can be covered by all the pieces.

- The blue cell is empty. Then there are nonempty red cells and we compute that the $2n(2n + 1)$ pieces contain $2n(2n + 1) \cdot 2 + 6n \cdot n = 14n^2 + 4n$ colored cells.
- One red cell is empty. Then there exists a piece S of type 1 that contains $n + 1$ red cells and $4n^2 + 2n - 1$ pieces of type 2. Altogether they contain $5n - 2$ red cells. Then the pieces contain $2n + 2 + (4n^2 + 2n - 1) \cdot 2 + (5n - 2)n = 13n^2 + 4n$ colored cells.
- The blue and red cells are not empty. Then the pieces of soap will contain n more colored cells than in the previous case.

However, the total number of colored cells within the cube is $1 + 6n + 12n^2$, and for $n \geq 3$, we have the inequality $13n^2 + 4n > 12n^2 + 6n + 1$, which is a contradiction, because the pieces are supposed to be nonoverlapping.

4. Second proof when $n > 3$. For $k \in \{1, 2, 3\}$, let us denote by m_k the number of pieces whose longest side is orthogonal to the plane of the slice $P_{k,i}$. Each such piece has exactly 2 cells in common with $P_{k,n+1}$. Every piece from the remaining ones has either $n + 1$ or $2(n + 1)$ cells in common with $P_{k,n+1}$. Therefore, $n + 1$ should divide the number of remaining cells, which is $4n(n + 1) - 2m_k = (2n + 1)^2 - 1 - 2m_k$. This implies that $n + 1$ divides m_k for all k.

Moreover, the total number of cells is $m_1 + m_2 + m_3 = 2n(2n + 1)$. Since $\gcd(n, n + 1) = \gcd(n, 2n + 1) = 1$, we derive that $n \in \{1, 3\}$.

Problem 2.65. Let F be a finite subset of $\mathbb{R}$ with the property that any value of the distance between two points from F (except for the largest one) is attained at least twice, i.e., for two distinct pairs of points. Prove that the ratio of any two distances between points of F is a rational number.

Solution 2.65. Let $F = \{s_1, \ldots, s_n\}$ be a finite set from a vector space V over $\mathbb{Q}$. Consider the $n(n - 1)/2$ difference vectors $s_i - s_j$, where $i < j$. Some difference vectors appear several times in the difference vectors sequence, and we call them D vectors. We will prove a more general result: The vector space generated by those differences that are not D vectors coincides with the vector space generated by all difference vectors.

Proof of the claim. Let us assume the contrary. By the Hahn–Banach lemma, there exists a linear functional $f : V \to \mathbb{Q}$ that vanishes on those vectors that are not D vectors, although f is nonzero on all difference vectors. Observe that we can replace V with a finite-dimensional linear subspace that contains F.

Let $M = f(s_i), m = f(s_j)$ be the greatest and respectively the smallest values from $f(F)$, where of course, $m \neq M$. The set of linear functionals $f : V \to \mathbb{Q}$ that map F one-to-one into $\mathbb{Q}$ is dense in the set of linear functionals. Consequently, there exists $g : V \to \mathbb{Q}$ that injects F into $\mathbb{Q}$ such that

$$|f(s) - g(s)| \leq \frac{1}{5}(M - m), \text{ for all } s \in F.$$

Consider $g(s_p)$ the maximal value and $g(s_q)$ the minimal value within the set $g(F) \subset \mathbb{Q}$. Therefore $s_p - s_q$ is not a D vector and thus f vanishes: $f(s_p - s_q) = 0$. This implies that

$$g(s_i) - g(s_j) \leq g(s_p) - g(s_q) \leq f(s_p) - f(s_q) + \frac{2}{5}(M - m) = \frac{2}{5}(M - m).$$

On the other hand, g is closely approximated by f; hence

$$g(s_i) - g(s_j) \leq g(s_p) - g(s_q) \leq f(s_p) - f(s_q) + \frac{2}{5}M - n = \frac{2}{5}(M - n),$$

and also

$$g(s_i) - g(s_j) \geq f(s_i) - \frac{1}{5}(M - m) - \left(f(s_j) + \frac{1}{5}(M - m)\right) = \frac{3}{5}(M - m),$$

contradicting the previous equality. This proves our general claim.

Assume now that $F \subset \mathbb{R}$, and the extreme points of F are 0 and 1. Consider then $V = \mathbb{R}$ as a vector space over $\mathbb{Q}$. From the hypothesis, 1 is the only distance that is not attained twice. The claim above tells us that the $\mathbb{Q}$-linear space generated by 1 contains all difference vectors. Thus all distances are rational.

Comments 53 *The Hahn–Banach lemma, alluded to above, says that for any vector subspace W of the vector space V and any vector $v \in V$ that does not belong to W, there exists a linear map $f : V \to \mathbb{R}$ such that $f(W) = 0$ but $f(v) \neq 0$.*

The result from the statement is due to Mikusiński and Schinzel, and the proof given above is due to Straus.

- J. Mikusiński, A. Schinzel: *Sur la réductibilité de certains trinômes,* Acta Arith. 9 (1964), 91–95.
- E.G. Straus: *Rational dependence in finite sets of numbers,* Acta Arith. 11 (1965), 203–204.

7

Geometry Solutions

7.1 Synthetic Geometry

Problem 3.1. Let I be the center of the circle inscribed in the triangle ABC and consider the points α, β, γ situated on the perpendiculars from I on the sides of the triangle ABC such that

$$|I\alpha| = |I\beta| = |I\gamma|.$$

Prove that the lines $A\alpha$, $B\beta$, $C\gamma$ are concurrent.

Solution 3.1. The lines $A\alpha$, $B\beta$, $C\gamma$ are concurrent if and only if they satisfy Ceva's theorem. Let us draw $A_1 A_2 \parallel BC$, where $A_1 \in |AB|$, $A_2 \in |AC|$, and $\alpha \in |A_1 A_2|$. We also draw their analogues. Using Thales' theorem, the Ceva condition is reduced to

$$\prod_{\text{cyclic}} \frac{|\alpha A_2|}{|\alpha A_1|} = 1.$$

Now I is situated on all three bisectors and $|I\gamma| = |I\alpha|$. By symmetry we have $|C_1\gamma| = |A_2\alpha|$ and their analogues. This shows that the Ceva condition is satisfied.

Problem 3.2. We consider the angle xOy and a point $A \in Ox$. Let (C) be an arbitrary circle that is tangent to Ox and Oy at the points H and D, respectively. Set AE for the tangent line drawn from A to the circle (C) that is different from AH. Show that the line DE passes through a fixed point that is independent of the circle (C) chosen above.

Solution 3.2. Let A' be the point on Oy such that $|OA'| = |OA|$. Let us further consider the intersection points $AE \cap Oy = \{F\}$ and $DE \cap |AA'| = \{P\}$.

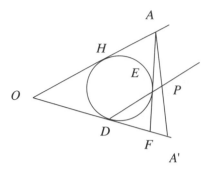

We use Menelaus's theorem in the triangle AFA' with the line DEP as transversal. We have

$$\frac{|PA'|}{|PA|} \cdot \frac{|EA|}{|EF|} \cdot \frac{|DF|}{|DA'|} = 1.$$

But one knows that $|EA| = |AH|$, $|EF| = |DF|$, and $|A'D| = |AH|$. Then $\frac{|PA'|}{|PA|} = 1$, and therefore P is the midpoint of $|AA'|$. Consequently, P is a fixed point.

Problem 3.3. Let C be a circle of center O and A a fixed point in the plane. For any point $P \in C$, let M denote the intersection of the bisector of the angle $\widehat{AOP}$ with the circle circumscribed about the triangle AOP. Find the geometric locus of M as P runs over the circle C.

Solution 3.3. One knows that OM is the bisector of $\widehat{AOP}$, and therefore OM is the mediator of the segment PB. Also, M is the midpoint of the arc of the circle $\widehat{AP}$, and therefore M belongs to the mediator of the segment $|AP|$. In the triangle PAB, M is the intersection of two mediators, and hence it also belongs to the mediator of segment $|AB|$. It is immediate now that the geometric locus of M is the entire mediator.

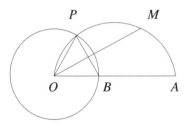

Problem 3.4. Let ABC be an isosceles triangle having $|AB| = |AC|$. If AS is an interior Cevian that intersects the circle circumscribed about ABC at S, then describe the geometric locus of the center of the circle circumscribed about the triangle BST, where $\{T\} = AS \cap BC$.

Solution 3.4. We have $\widehat{BST} = \widehat{BCA} = \widehat{CBA}$. Thus the line AB is tangent to the circle circumscribed about BST. Consequently, the center of this circle is situated on the line that passes through B that is perpendicular to AB.

Problem 3.5. Let AB, CD, EF be three chords of length one on the unit circle. Then the midpoints of the segments $|BC|$, $|DE|$, and $|AF|$ form an equilateral triangle.

Solution 3.5. Let a, b, c, d, e, f be the complex affixes (coordinates) of the respective points on the unit circle and $\varepsilon = \exp\left(\frac{2\pi i}{3}\right)$. Then the triangles OAB, OCD, OEF are equilateral, and therefore we have the following complex identities:

$$a + \epsilon b = 0,$$
$$\epsilon c + \epsilon^2 d = 0,$$
$$\epsilon^2 e + f = 0.$$

These imply that $\frac{a+f}{2} + \epsilon \frac{b+c}{2} + \epsilon^2 \frac{d+c}{2} = 0$. Now the midpoint of $|AF|$ has the affix $\frac{a+f}{2}$, and the other midpoints are similarly associated to $\frac{b+c}{2}$ and $\frac{d+e}{2}$. The relation above is equivalent to the claim of the statement.

Problem 3.6. Denote by P the set of points of the plane. Let $\star : P \times P \to P$ be the following binary operation: $A \star B = C$, where C is the unique point in the plane such that ABC is an oriented equilateral triangle whose orientation is counterclockwise. Show that $\star$ is a nonassociative and noncommutative operation satisfying the following "medial property":

$$(A \star B) \star (C \star D) = (A \star C) \star (B \star D).$$

Solution 3.6. Let $A \neq B$, $A \star B = C$. Then $(A \star B) \star C = C \star C = C$. But $A \star (B \star C) = A \star A = A \neq C$; therefore the operation $\star$ is nonassociative. Because $A \star B$, $B \star A$ are symmetric with respect to AB, the operation $\star$ is noncommutative.

Now identify the points of the plane with complex numbers. Set $\varepsilon = \exp\left(\frac{2\pi i}{3}\right)$. We have then $A \star B = A + \varepsilon(B - A)$. It is now easy to verify that

$$(A \star B) \star (C \star D) = (1 - \varepsilon)^2 A + \varepsilon(1 - \varepsilon)(B + C) + \varepsilon^2 D.$$

Moreover, this expression is symmetric in B and C.

Comments 54 *It is possible to take an arbitrary $\varepsilon \in \mathbb{C}$ and to define $A \star_\varepsilon B = C$, where C is the point such that the triangle ABC is similar to the triangle determined by the points 0, 1, and ε. Again, the medial property holds for this binary law.*

Problem 3.7. Consider two distinct circles C_1 and C_2 with nonempty intersection and let A be a point of intersection. Let P, $R \in C_1$ and Q, $S \in C_2$ be such that PQ and RS are the two common tangents. Let U and V denote the midpoints of the chords PR and QS. Prove that the triangle AUV is isosceles.

Solution 3.7. Let B be the other intersection point of the two circles. It is known that the line AB is the radical axis of the two circles, i.e., the geometric locus of the points having the same power with respect to the two circles. Let us then consider the intersection point $AB \cap |PQ| = X$. The power of X with respect to either circle is $P_{C_1}(X) = |XA| \cdot |XB|$. On the other hand, $P_{C_1}(X) = |XP|^2$ and $P_{C_2}(X) = |XQ|^2$, and so we obtain that X is the midpoint of $|PQ|$. Thus the line AB passes through the midpoints of $|PQ|$ and $|RS|$, and thus it is also passes through the midpoint of $|UV|$, being also orthogonal to the latter. This shows that AUV is isosceles.

Problem 3.8. If the planar triangles AUV, VBU, and UVC are directly similar to a given triangle, then so is ABC. Recall that two triangles are directly similar if one can obtain one from the other using a homothety with positive ratio, rotations and translations.

Solution 3.8. Using complex numbers, each similarity (i.e., homothety) has the form $z \mapsto \alpha_i z + \beta_i$, for some $\alpha_i, \beta_i \in \mathbb{C}$. The fixed triangle has its vertices located at z_1, z_2, and z_3, and the points A, B, C, U, V are located at a, b, c, u, v respectively. By hypothesis, $u = \alpha_i z_{i+1} + \beta_i$ and $v = \alpha_i z_{i+2} + \beta_i$, where the indices are considered mod 3. This implies that $\alpha_i z_i (z_{i+1} - z_{i+2}) = z_i(u - v)$ and $\beta_i(z_{i+1} - z_{i+2}) = z_{i+1}v - z_{i+2}u$, so that

$$\sum_{i=1}^{3} (\alpha_i z_i + \beta_i)(z_{i+1} - z_{i+2}) = 0.$$

But from

$$\sum_{i=1}^{3} (z_{i+1} - z_{i+2}) = 0, \quad \sum_{i=1}^{3} z_i (z_{i+1} - z_{i+2}) = 0$$

we derive

$$\det \begin{pmatrix} a & z_1 & 1 \\ b & z_2 & 1 \\ c & z_3 & 1 \end{pmatrix} = \det \begin{pmatrix} \alpha_1 z_1 + \beta_1 & z_1 & 1 \\ \alpha_2 z_2 + \beta_2 & z_2 & 1 \\ \alpha_3 z_3 + \beta_3 & z_3 & 1 \end{pmatrix} = 0,$$

which is equivalent to saying that ABC is similar to the fixed triangle.

Problem 3.9. Find, using a straightedge and a compass, the directrix and the focus of a parabola. Recall that the parabola is the geometric locus of those points P in the plane that are at equal distance from a point O (called the focus) and a line d called the directrix.

Solution 3.9. Let Δ be an arbitrary direction in the plane. We draw two distinct lines Δ_1 and Δ_2 cutting the parabola in two nontrivial chords $X_1 X_2$ and $Y_1 Y_2$, both being parallel to Δ. Let X, Y be the midpoints of $|X_1 X_2|$ and respectively $|Y_1 Y_2|$. It is simple to show that the line XY (called the diameter conjugate to the direction Δ) is parallel to the symmetry axis of the parabola. Moreover, let T_Δ be the intersection point between the line XY and the parabola. Then the line that passes through T_Δ and is parallel to Δ is tangent (at T_Δ) to the parabola.

Now we will use a well-known theorem in the geometry of conics, which claims that if P is a point in the plane (outside the convex hull of the parabola), PA, PB are the two tangents to the parabola issued from P (where A and B belong to the parabola) and if the angle $\widehat{APB} = \frac{\pi}{2}$, then AB is a focal chord, i.e., the focus F lies on the chord $|AB|$.

Choose now the direction $\Delta^{\perp}$ that is orthogonal to Δ and construct as above the point $T_{\Delta^{\perp}}$ on the parabola. Then, according to the previous result, the chord $T_{\Delta}T_{\Delta^{\perp}}$ is focal.

Choose next another couple of orthogonal directions, δ and $\delta^{\perp}$, and derive the new focal chord $T_{\delta}T_{\delta^{\perp}}$. Then the two focal chords intersect at the focus $T_{\Delta}T_{\Delta^{\perp}} \cap T_{\delta}T_{\delta^{\perp}} = \{O\}$. Moreover, we can draw the axis of symmetry of the parabola as the line passing through the focus that is parallel to (any) diameter conjugate to XY.

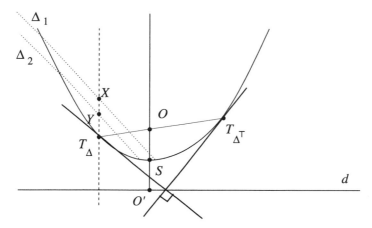

Finally, let S be the intersection of the axis with the parabola and let O' be the symmetric of the focus with respect to S. Then the directrix d is the line passing through O' that is orthogonal to the axis OSO'.

Then $\tilde{\Delta}$, the conjugate of Δ, meets Δ at a point on the axis. We represent then Δ, $\tilde{\Delta}$ and we find two points of the axis.

Problem 3.10. Prove that if M is a point in the interior of a circle and $AB \perp CD$ are two chords perpendicular at M, then it is possible to construct an inscribable quadrilateral with the following lengths:

$$\big||AM| - |MB|\big|, \quad |AM| + |MB|, \quad \big||DM| - |MC|\big|, \quad |DM| + |MC|.$$

Solution 3.10. Let us construct $AF \parallel CD$ and $CE \parallel AB$, where the points E, F lie on the given circle.

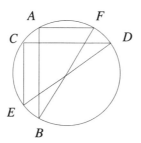

Then the intersection point $BF \cap ED = O$ is the center of the circle. Moreover, we can compute $|CE| = \big||MB| - |MA|\big|$ and $|AF| = \big||MD| - |MC|\big|$. We rotate the triangle CED around the center O in order to superpose ED onto FB. Call C' the new position of the point C. Then, the length of the fourth side of the quadrilateral $ABC'F$ is $|BC'| = |DM| + |MC|$.

Problem 3.11. If the Euler line of a triangle passes through the Fermat point, then the triangle is isosceles.

Solution 3.11. The Euler line OG is determined by the circumcenter O and the center of gravity G. The trilinear coordinates of the points involved are $G(\sin B \sin C, \sin A \sin C, \sin A \sin B)$, $O(\cos A, \cos B, \cos C)$, and the Fermat point $F(\sin(B + 60), \sin(C + 60), \sin(A + 60))$. Therefore, G, O, and F are collinear only if the determinant of their trilinear coordinates vanishes. By direct calculation, this determinant can be computed as

$$\sin(A - B)\sin(B - C)\sin(C - A) = 0,$$

and hence the triangle must be isosceles.

Problem 3.12. Consider a point M in the interior of the triangle ABC, and choose $A' \in AM$, $B' \in BM$, and $C' \in CM$. Let P, Q, R, S, T, and U be the intersections of the sides of ABC and $A'B'C'$. Show that PS, TQ, and RU meet at M.

Solution 3.12. The triangles ABC and $A'B'C'$ are in perspective with center M, and hence $A'C' \cap AC$, $B'C' \cap BC$, and $B'A' \cap BA$ are collinear. Let xy be the line on which these points lie.

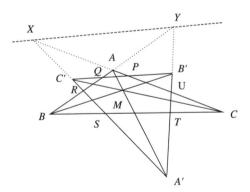

We make a projective transformation that sends the line xy into the line at infinity. This way, the sides of the two triangles become pairwise parallel.

Consider A' fixed and let the points B'', C'' be mobile points sitting on AB' and AC', respectively. If we move the line $B''C''$ by keeping it parallel to BC, then the intersection points $Q' = AB \cap B''C''$ and $P' = AC \cap B''C''$ determine homographic divisions on AB and AC. When we reach the position where $A \in B''C''$, we see that $Q'T \cap SP' = A$; also, when we reach the position for which $A' \in B''C''$, we find that $Q'T \cap SP' = A'$. Therefore, since Q', P' determine homographic divisions, we obtain that for any position of $B''C''$, we have $Q'T \cap SP' \in AA'$. In particular, this happens when $B''C'' = B'C'$, and thus $QT \cap SP \in AA'$. Analogously, we have $QT \cap RU \in CC'$, $SP \cap RU \in BB'$. According to Pascal's theorem, the diagonals of the hexagon $PQRSTU$ are concurrent, since the edges are pairwise parallel. This shows that $QT \cap SP \cap RU \in AA' \cap BB' \cap CC' = M$.

Problem 3.13. Show that if an altitude in a tetrahedron crosses two other altitudes, then all four altitudes are concurrent.

Solution 3.13. If the altitude h_A from A intersects h_B, then $AB \perp CD$. Similarly, if h_A intersects h_C, then $AC \perp BD$. These can be written vectorially as $\langle \overrightarrow{AB}, \overrightarrow{AD} - \overrightarrow{AC} \rangle = 0$, and $\langle \overrightarrow{AC}, \overrightarrow{AD} - \overrightarrow{AB} \rangle = 0$, where $\langle , \rangle$ denotes the Euclidean scalar product. These imply that $\langle \overrightarrow{AB} - \overrightarrow{AC}, \overrightarrow{AD} \rangle = 0$, which is equivalent to $AD \perp BC$. This implies that all four altitudes are concurrent.

Problem 3.14. Three concurrent Cevians in the interior of the triangle ABC meet the corresponding opposite sides at A_1, B_1, C_1. Show that their common intersection point is uniquely determined if $|BA_1|$, $|CB_1|$, and $|AC_1|$ are equal.

Solution 3.14. Let $|BA_1| = |CB_1| = |AC_1| = \delta$. From Ceva's theorem, we obtain

$$(a - \delta)(b - \delta)(c - \delta) = \delta^3.$$

We may assume that $a \leq b \leq c$, and hence $\delta \in [0, a]$.

The left-hand side is a decreasing function of δ ranging from the value abc, obtained for for $\delta = 0$, to 0, when $\delta = a$. Further, the right-hand-side function δ^3 is strictly increasing. Therefore, the two functions can have only one common value on the interval $[0, a]$. This determines uniquely the intersection point.

Comments 55 *Also, if we ask that $|CA_1|$, $|BC_1|$, and $|AB_1|$ be equal, their common value $\tilde{\delta}$ coincides with the δ we found above. In particular, the two intersection points are isotomic to each other.*

Problem 3.15. Let $ABCD$ be a convex quadrilateral with the property that the circle of diameter AB is tangent to the line CD. Prove that the circle of diameter CD is tangent to the line AB if and only if AD is parallel to BC.

Solution 3.15. Let M be the midpoint of $|AB|$ and draw the perpendicular $MM' \perp CD$, where $M' \in CD$. Then M' is the contact point of the circle C_1 of diameter AB and the line CD. Thus $\widehat{AM'B} = \frac{\pi}{2}$, and so $|MM'| = \frac{|AB|}{2}$. Conversely, if M, M' are as above and satisfy the condition $|MM'| = \frac{|AB|}{2}$, then the circle of diameter AB is tangent to CD.

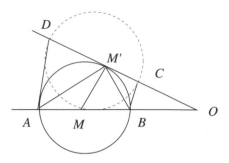

Let now $O = AB \cap CD$, assuming that the two lines are not parallel. If $|OM| = \lambda$ and $\widehat{AOD} = \theta$, then $|MM'| = \lambda \sin \theta$ and so $|MB| = |MA| = \lambda \sin \theta$. This turns to

$$\frac{|OB|}{|OA|} = \frac{|OM| + |MA|}{|OM| - |MB|} = \frac{\lambda(1 + \sin \theta)}{\lambda(1 - \sin \theta)} = \frac{1 + \sin \theta}{1 - \sin \theta}.$$

Conversely, $\frac{|OB|}{|OA|} = \frac{1 + \sin \theta}{1 - \sin \theta}$ implies that $|MM'| = |MA| = |MB|$.

Let now P be the midpoint of CD and draw the perpendicular $PP' \perp AB$, with $P' \in AB$. The previous argument shows that $|PP'| = \frac{|CD|}{2}$ is equivalent to $\frac{|OC|}{|OD|} = \frac{1 - \sin \theta}{1 + \sin \theta}$, which, according to Thales' theorem, is equivalent to saying that AD is parallel to BC. But $|PP'| = \frac{|CD|}{2}$ iff the circle of diameter CD is tangent to AB, as claimed.

The last case occurs when AB is parallel to CD. Then $|MM'|$ is the distance between the lines AB and CD, and hence $|MM'| = \frac{|AB|}{2} = \frac{|CD|}{2}$. Thus $ABCD$ is a parallelogram, and hence AD is parallel to BC.

Problem 3.16. Let A', B', C' be points on the sides BC, CA, AB of the triangle ABC. Let M_1, M_2 be the intersections of the circle $A'B'C'$ with the circle ABA' and let N_1, N_2 be the analogous intersections of the circle $A'B'C'$ with the circle ABB'.

1. Prove that $M_1 M_2$, $N_1 N_2$, AB are either parallel or concurrent, in a point that we denote by A_1';
2. Prove that the analogously defined points A_1', B_1', C_1' are collinear.

Solution 3.16. 1. More precisely, $M_1 M_2$, $N_1 N_2$, AB are the radical axes of the three pairs of circles formed by $A'B'C'$, ABB', and ABC'. The first part follows.

2. Further, A_1' is the radical center of $A'B'C'$, ABB', ABA', and thus $|A_1' A| \cdot |A_1' B| = p_{ABB'}^{A_1'} = p_{A'B'B'}^{A_1'}$, where p denotes the power of the point with respect to the given circle. Therefore $p_{A'B'C'}^{A_1'} = p_{ABC}^{A_1'}$, and their analogues, which imply that the three points A_1', B_1', and C_1' belong to the radical axis of the pair of circles $A'B'C'$ and ABC.

Problem 3.17. A circumscribable quadrilateral of area $S = \sqrt{abcd}$ is inscribable.

Solution 3.17. By the circumscribability hypothesis, we have $a + c = b + d$. Let k be the length of the diagonal that leaves a, b on one side and c, d on the other side.

Let α, β be the angles of the quadrilateral between a and b, and c and d, respectively. We have then the law of cosines computing k:

$$k^2 = a^2 + b^2 - 2ab \cos \alpha,$$
$$k^2 = c^2 + d^2 - 2cd \cos \beta.$$

Subtracting the terms $(a - b)^2 = (c - d)^2$, we obtain

$$2ab(1 - \cos \alpha) = 2cd(1 - \cos \beta).$$

Now compute the area S of the quadrilateral $S = \frac{1}{2}(ab \sin \alpha + cd \sin \beta)$, which is by hypothesis $\sqrt{abcd}$. Therefore

$$4S^2 = 4abcd = a^2b^2(1 - \cos^2 \alpha) + c^2d^2(1 - \cos^2 \beta) + 2abcd \sin \alpha \sin \beta.$$

From the former identity, we obtain

$$4abcd = ab(1+\cos \alpha)cd(1-\cos \beta)+cd(1+\cos \beta)ab(1-\cos \alpha)+2abcd \sin \alpha \sin \beta.$$

After simplification of the terms, we obtain

$$4 = 2 - 2\cos(\alpha + \beta),$$

and therefore $\alpha + \beta = \pi$. This implies that the quadrilateral is inscribable.

Problem 3.18. Let O be the center of the circumcircle, Ge the Gergonne point, Na the Nagel point, and G_1, N_1, the isogonal conjugates of G and N, respectively. Prove that G_1, N_1, and O are collinear (see also the Glossary for definitions of the important points in a triangle).

Solution 3.18. We will use the trilinear coordinates with respect to the given triangle. These are triples of numbers (see the triangle compendium from the Glossary) proportional to the distances from the given point to the edges of the triangle. The trilinear coordinates behave very nicely when we consider the isogonal conjugate. If P has the coordinates (x, y, z), then its isogonal conjugate has coordinates $\left(\frac{1}{x}, \frac{1}{y}, \frac{1}{z}\right)$.

The trilinear coordinates of the points considered above are then

$$O(\cos A, \cos B, \cos C),$$

$$Ge\left(\frac{1}{a(p-a)}, \frac{1}{b(p-b)}, \frac{1}{c(p-c)}\right), \quad G_1\left(a(p-a), b(p-b), c(p-c)\right),$$

$$Na\left(\frac{p-a}{a}, \frac{p-b}{b}, \frac{p-c}{c}\right), \quad N_1\left(\frac{a}{p-a}, \frac{b}{p-b}, \frac{c}{p-c}\right).$$

Further, three points are collinear if the determinant of the 3×3 matrix of their coordinates vanishes. The claim from the statement is then equivalent to the following identity:

$$\det \begin{pmatrix} \cos A & \cos B & \cos C \\ a(p-a) & b(p-b) & c(p-c) \\ \frac{a}{p-a} & \frac{b}{p-b} & \frac{c}{p-c} \end{pmatrix} = 0.$$

We develop the determinant and obtain

$$\sum_{\text{cyclic}} \cos A \ bc \left(\frac{p-b}{p-c} - \frac{p-c}{p-b} \right) = \sum_{\text{cyclic}} \cos A \ \frac{bc}{(p-b)(p-c)} \ a(c-b).$$

This vanishes iff

$$\sum_{\text{cyclic}} \cos A \ (p-a)(c-b) = 0,$$

which is equivalent to

$$\sum_{\text{cyclic}} (b^2 + c^2 - a^2)(c-b)(-a+b+c)a = 0,$$

which is immediate.

Note that we could use instead the barycentric coordinates as well. These coordinates are triples of numbers proportional to the areas of the three triangles determined by the point and two of the vertices of the initial triangle. The barycentric coordinates of the points considered above are given by

$$O(a,b,c), \quad Ge\left(\frac{1}{p-a}, \frac{1}{p-b}, \frac{1}{p-c} \right), \quad G_1\left(a^2(p-a), b^2(p-b), c^2(p-c) \right),$$

$$Na\,(p-a, p-b, p-c), \quad N_1\left(\frac{a^2}{p-a}, \frac{b^2}{p-b}, \frac{c^2}{p-c} \right).$$

Finally, three points are collinear if and only if their coordinate matrix has vanishing determinant.

7.2 Combinatorial Geometry

Problem 3.19. Consider a rectangular sheet of paper. Prove that given any $\varepsilon > 0$, one can use finitely many foldings of the paper along its sides in either 2 equal parts or 3 equal parts to obtain a rectangle whose sides are in ratio r for some r satisfying $1 - \epsilon \le r \le 1 + \epsilon$.

Solution 3.19. Observe that $\log_3 2 \notin \mathbb{Q}$ and thus the residues modulo 1 of the elements $n \log_3 2$, with $n \in \mathbb{Z}_+$, form a dense subset of $(0, 1)$. This implies that the set of numbers of the form $2^a 3^b$, with $a, b \in \mathbb{Z}$, are dense in $\mathbb{R}$.

Problem 3.20. Show that there exist at most three points on the unit disk with the distance between any two being greater than $\sqrt{2}$.

Solution 3.20. Consider the points A_i on the unit disk of radius 1 and center O. Then $|OA_i| \le 1$, and if we assume that $|A_i A_j| > \sqrt{2}$, then $2 < |OA_i|^2 + |OA_j|^2 - 2|OA_i| \cdot |OA_j| \cos \widehat{A_i O A_j}$. Thus, $\cos \widehat{A_i O A_j} < 0$ and hence $\widehat{A_i O A_j} > \frac{\pi}{2}$, which implies that $n \le 3$.

Comments 56 *H.S.M. Coxeter proposed this problem in 1933. The same argument shows that if we have k points A_i within the unit disk, then there exists a pair of points such that $|A_i A_j| \le \max\left(1, 2 \sin \frac{\pi}{k}\right)$, and this is sharp for $2 \le k \le 7$.*

The result was generalized to higher dimensions by Davenport, Hajós, and, independently, Rankin, as follows: There are at most $n + 1$ points in the unit ball of $\mathbb{R}^n$ such that the distance between any two points is greater than $\sqrt{2}$.

Here is a proof sketch. Let A_i denote the m points satisfying the hypothesis, so that they are different from the origin O. Then we have $m \ge n + 2$ unit vectors $\frac{OA_i}{|OA_i|}$ in $\mathbb{R}^n$ with the property that $\langle v_i, v_j \rangle < 0$. We claim next that there exist pairwise orthogonal subspaces $W_1, W_2, \dots, W_{m-n} \subset \mathbb{R}^n$ that span $\mathbb{R}^n$ such that each space W_j contains $1 + \dim W_j$ vectors among the unit vectors above, $v_1, \dots, v_m$, that span W_j. Since $m - n \ge 2$, there exist two vectors among the v_i that are orthogonal to each other, which contradicts our assumptions. The claim follows by induction on the dimension n. If $n = 2$, it is immediate. Now, if $v_i = -v_n$, for some $i < n$, then all other v_j should be orthogonal to v_n (and to v_i). Otherwise, we can write for all $i < n$, $v_i = u_i + x_i v_n$, for some scalars $x_i < 0$ and some nonzero vector u_i orthogonal to v_n. This implies that $\langle u_i, u_j \rangle < 0$, when $i \ne j$. We can apply therefore the induction hypothesis for the subspace orthogonal to v_n and the set of vectors v_i, $i < n$, and obtain the subspaces W_j'. We define then $W_i = W_i'$, for $i \ge 2$, and let W_1 be the span of W_1' and v_n. This proves the claim.

The claim shows that if we have $m \ge n + 2$ points on the unit sphere in $\mathbb{R}^n$ whose pairwise distances are at least $\sqrt{2}$, then $m \le 2n$, and moreover, there exist two points at distance precisely $\sqrt{2}$.

- H.S.M. Coxeter: Amer. Math. Monthly 40 (1933), 192–193.
- R.A. Rankin: *The closest packing of spherical caps in n dimensions*, Proc. Glasgow Math.Assoc. 2 (1955), 139–144.

Problem 3.21. A convex polygon with $2n$ sides has at least n diagonals not parallel to any of its sides. The equality is attained for the regular polygon.

Solution 3.21. Let l be a side of the polygon. Then we have at most $n - 2$ diagonals parallel to l. In fact, let $d_1, d_2, \dots, d_p$ be these diagonals. Then the pair of segments (d_1, l) determines a quadrilateral (actually a trapezoid) having three common sides with the polygon. Moreover, all the pairs $(d_2, d_1), \dots, (d_{p-1}, d_p)$ determine quadrilaterals, each one containing two sides of the initial polygon. This implies that $p \le n - 2$.

The total number of diagonals that are parallel to at least one side is at most $2n(n - 2)$. But there are $\frac{2n(2n-3)}{2}$ diagonals, and hence $n(2n - 3) - 2n(n - 2) = n$ diagonals with the desired property.

Problem 3.22. Let d be the sum of the lengths of the diagonals of a convex polygon $P_1 \cdots P_n$ and p its perimeter. Prove that for $n \geq 4$, we have

$$n - 3 < 2\frac{d}{p} < \left[\frac{n}{2}\right]\left[\frac{n+1}{2}\right] - 2.$$

Solution 3.22. Set $\alpha_k = \sum_{i=1}^{n} |P_i P_{i+k}|$. We will prove first that

$$p = \alpha_1 < \alpha_2 < \alpha_3 < \cdots < \alpha_{\left[\frac{n}{2}\right]} \quad \text{and} \quad \alpha_{\left[\frac{n}{2}\right]+1} > \cdots > \alpha_{n-2} > \alpha_{n-1} = p.$$

Let us show first that $p < \alpha_2$. Let $A_i A_{i+2} \cap A_{i+1} A_{i+3} = \{O_{i+1}\}$. Observe that $|A_{i+1} O_{i+1}| \leq |A_{i+1} O_{i+2}|$, since $A_{i+1}\overset{\frown}{A_{i+2}}O_{i+1} \leq A_{i+1}\overset{\frown}{A_{i+2}}O_{i+2}$. The triangle inequality in $A_i O_i A_{i+1}$ further yields

$$|A_{i+1} O_{i+1}| + |O_{i+1} A_{i+2}| > |A_{i+1} A_{i+2}|.$$

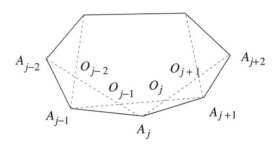

Summing up these inequalities, and using the remark, it follows that $p < \alpha_2$, as claimed.

Let now $k < \left[\frac{n}{2}\right] - 1$, and set $A_i A_{i+k} \cap A_{i+1} A_{i+k+1} = \{O\}$. The triangle inequality again shows that

$$|A_i A_{i+k}| + |A_{i+1} A_{i+k+1}| = |A_i O| + |O A_{i+k}| + |A_{i+1} O| + |O A_{i+k+1}|$$
$$> |A_{i+1} A_{i+k}| + |A_i A_{i+k+1}|.$$

By summing up over i, we obtain

$$2\alpha_k > \alpha_{k+1} + \alpha_{k-1};$$

thus

$$\alpha_{k+1} - \alpha_k < \alpha_k - \alpha_{k-1} < \cdots < \alpha_2 - \alpha_1,$$

and hence the claim.

Moreover, if $n = 2m$, then

$$2\alpha_m > \alpha_{m+1} + \alpha_{m-1} = 2\alpha_{m+1},$$

and so $\alpha_m > \alpha_{m-1}$. Thus, we have proved that $\alpha_{k+1} > \alpha_k$ for all k. By summing up all these inequalities over k, we find that $2d > (n-3)p$.

Further, consider the triangle inequality $|A_i A_j| < |A_i A_{i+1}| + \cdots + |A_{j-1} A_j|$ and sum up over all i and j; this yields

$$2d < \left(\left(\left[\frac{n}{2} \right] \left[\frac{n+1}{2} \right] \right) - 2 \right) p.$$

Problem 3.23. Find the convex polygons with the property that the function $D(p)$, which is the sum of the distances from an interior point p to the sides of the polygon, does not depend on p.

Solution 3.23. Let K be our polygon, p an interior point, $a_1, \ldots, a_n$ the distances from p to the sides of K, and let $u_1, \ldots, u_n$ be unit vectors perpendicular to the sides. Choose another point, say q, in the interior of K. Denote by $\bar{v}$ the vector $\overrightarrow{pq}$. The distance from q to the side with label i is therefore $d_i = a_i + \langle u_i, v \rangle$, where $\langle , \rangle$ is the scalar product. It follows that

$$D(q) = D(p) + \left\langle \sum_{i=1}^{n} u_i, v \right\rangle.$$

Therefore, a necessary and sufficient condition for $D(p)$ to be constant is that

$$\sum_{i=1}^{n} u_i = 0.$$

Observe that the solution can be extended to convex polyhedra in $\mathbb{R}^m$.

Problem 3.24. Prove that a sphere of diameter 1 cannot be covered by n strips of width l_i if $\sum_{i=1}^{n} l_i < 1$. Prove that a circle of diameter 1 cannot be covered by n strips of width l_i if $\sum_{i=1}^{n} l_i < 1$.

Solution 3.24. 1. The lateral area of the body obtained by cutting off the sphere with two planes at distance h is $2\pi R h$. Let us assume that we have a covering by means of the strips B_i. Then the sum of areas cut open by the strips B_i is at least the area of the sphere, hence $\pi \sum_{i=1}^{n} l_i \geq \pi$, contradicting our assumptions.
   2. If we had such a covering, let B_i^* be the spatial strip that cuts the planar strip B_i, orthogonal to the plane of the circle. Then the union of the strips B_i^* would cover the cylinder over the disk, in particular the sphere of diameter 1. This would contradict the first part.

Problem 3.25. Consider n points lying on the unit sphere. Prove that the sum of the squares of the lengths of all segments determined by the n points is less than n^2.

Solution 3.25. Let O be the center of the sphere and let X_i denote the given points. We denote by v_j the vector $\overrightarrow{OX_j}$, and by $\langle , \rangle$ the Euclidean inner product. We have then

$$|X_i X_j|^2 = \langle v_j - v_i, v_j - v_i \rangle.$$

Set $S = \sum_{i,j} |X_i X_j|^2$. We have then

$$2S = \sum_{i=1}^{n} \langle v_i - v_1, v_i - v_1 \rangle + \sum_{i=1}^{n} \langle v_i - v_2, v_i - v_2 \rangle + \cdots + \sum_{i=1}^{n} \langle v_i - v_n, v_i - v_n \rangle$$

$$= 2n(|v_1|^2 + \cdots + |v_n|^2) - 2 \sum_{i,j} \langle v_j, v_i \rangle$$

$$= 2n \sum_{i=1}^{n} |v_i|^2 - 2\langle v, v \rangle = 2n^2 - 2|v|^2 \le 2n^2,$$

where $v = \sum_{i=1}^{n} v_i$.

Problem 3.26. The sum of the vectors $\overrightarrow{OA_1}, \ldots, \overrightarrow{OA_n}$ is zero, and the sum of their lengths is d. Prove that the perimeter of the polygon $A_1 \ldots A_n$ is greater than $4d/n$.

Solution 3.26. If A is an arbitrary point in the plane, then $n\overrightarrow{OA} = \sum_{i=1}^{n}(\overrightarrow{OA_i} + \overrightarrow{A_i A}) = \sum_{i=1}^{n} \overrightarrow{A_i A}$, which implies that $n|OA| \le \sum |A_i A|$. Let us put $A = A_i$ for $i = 1, 2, \ldots, n$ and sum up the inequalities so obtained:

$$d = |OA_1| + \cdots + |OA_n| \le \frac{2}{n} \sum_{i,j} |A_i A_j|.$$

Using the triangle inequality,

$$|A_i A_j| < |A_i A_{i+1}| + \cdots + |A_{j-1} A_j|,$$

we infer that

$$d \le \frac{2}{n} \left(p + 2p + \cdots + \frac{n-1}{2} p \right) = \frac{n^2 - 1}{4n} p.$$

Therefore $p \ge 4d \frac{n}{n^2 - 1} > \frac{4d}{n}$.

Problem 3.27. Find the largest numbers a_k, for $1 \le k \le 7$, with the property that for any point P lying in the unit cube with vertices $A_1 \ldots A_8$, at least k among the distances $|PA_j|$ to the vertices are greater or equal than a_k.

Solution 3.27. At least eight distances are greater than or equal to 0, and this is sharp by taking $P = A_j$, and hence $a_8 = 0$.

By taking P to be the midpoint of an edge, we get $a_7 \le \frac{1}{2}$. Assume that $a_7 < \frac{1}{2}$. Then there exists some P for which at least two distances $|PA_j|$ are strictly smaller than $\frac{1}{2}$, which is obviously false, since vertices are distance 1 apart. Thus $a_7 = \frac{1}{2}$.

Consider P as the center of some face of the cube. We obtain then $a_6 \le a_5 \le \frac{\sqrt{2}}{2}$. If equality does not hold above, then there exists P for which three distances are smaller than $\frac{\sqrt{2}}{2}$. However, given three vertices of the cube, there are at least two of them at distance greater than or equal to the diagonal of a face, namely $\sqrt{2}$. Thus our assumption fails and so $a_6 = a_5 = \frac{\sqrt{2}}{2}$.

Now let P be the center of the cube. We derive that $a_4 \le a_3 \le a_2 \le a_1 \le \frac{\sqrt{3}}{2}$. Assume that $a_4 < \frac{\sqrt{3}}{2}$. Then there exist P and at least five distances strictly smaller than $\frac{\sqrt{3}}{2}$. However, for any choice of five vertices of a cube, one finds a pair of opposite vertices that are distance $\sqrt{3}$ apart. This contradicts our assumptions and thus $a_4 = a_3 = a_2 = a_1 = \frac{\sqrt{3}}{2}$.

Problem 3.28. The line determined by two points is said to be admissible if its slope is equal to 0, 1, −1, or ∞. What is the maximum number of admissible lines determined by n points in the plane?

Solution 3.28. We join two points by an edge if the line determined by them is admissible. We obtain a graph with n vertices. By hypothesis, each vertex has degree at most 4. Since the sum of degrees over all vertices is twice the number of edges, we find that there are at most $2n$ edges, thus at most $2n$ admissible lines. Moreover, there are only four directions in the plane, and the same computation shows that there are at most $\frac{1}{2}n$ edges associated to every direction. This implies that for odd n, the maximum number of admissible lines is $2n - 2$.

It is easy to construct examples in which we have precisely $2n$ admissible lines if n is even and $n \ge 12$. For odd n, the associated graph is obtained by adding an isolated vertex to a maximal graph of order $n - 1$, and this works for any $n \ge 3$.

Finally, if $f(n)$ denotes the maximum number of admissible lines, then explicit inspection shows that $f(2) = 1$, $f(3) = 3$, and $f(4) = 6$, and so the upper bound $2n$ is not attained. It can be shown that $f(6) = 11$ and $f(10) = 19$. Therefore, for all n but those from $\{2, 3, 4, 6, 10\}$, we have $f(n) = 4\left[\frac{n}{2}\right]$.

Comments 57 *More generally, assume that the admissible directions are k in number. The same proof shows that there therefore exist at most $\left[\frac{kn}{2}\right]$ admissible lines determined by n points. Moreover, the minimum number of directions (not necessarily admissible) defined by n points is at least $n - 1$. See also the article:*

- P.R. Scott: *On sets of direction determined by n points*, Amer. Math. Monthly 77 (1970), 502–505.

Problem 3.29. If $A = \{z_1, \ldots, z_n\} \subset \mathbb{C}$, then there exists a subset $B \subset A$ such that

$$\left|\sum_{z \in B} z\right| \ge \pi^{-1} \sum_{i=1}^n |z_i|.$$

Solution 3.29. We reorder the numbers $z_1, \ldots, z_n, -(z_1 + \cdots + z_n)$ in a sequence $w_1, \ldots, w_{n+1}$ such that the sequence arg $w_1, \ldots,$ arg w_{n+1} is monotonic. In the complex plane, the points $w_1, w_1 + w_2, \ldots, w_1 + \cdots + w_{n+1}$, are the vertices of a convex polygon P.

Set $d(P)$ for the diameter of P. Since the diameter of a polygon is the distance between two vertices, there exists $B \subset A$ such that

$$d(P) = \left|\sum_{z_i \in B} z_i\right|.$$

Let $p(P)$ be the perimeter of P. We have then

$$p(P) = \sum^n_i |w_i| = \sum^n_i |z_i| + \left| \sum^n_i z_i \right|.$$

The main isoperimetric inequality states that for any convex planar polygon Q we have the inequality $\pi d(Q) \geq p(Q)$. Applying this to the polygon P settles the claim.

Comments 58 *Using a refined isoperimetric inequality for polygons with a given number of vertices, one finds that there exists $B \subset \{1, \ldots, n\}$ such that*

$$\left| \sum_{i \in B} z_i \right| \geq \frac{\sin(k\pi/(2k+1))}{(2k+1)\sin(\pi/2k+1)} \sum^n_{i=1} |z_i|.$$

Problem 3.30. Let $A_1, \ldots, A_n$ be the vertices of a regular n-gon inscribed in a circle of center O. Let B be a point on the arc of circle $A_1 A_n$ and set $\theta = \widehat{A_n O B}$. If we set $a_k = |BA_k|$, then find the sum

$$\sum^n_{k=1} (-1)^k a_k$$

in terms of θ.

Solution 3.30. We have $a_k = 2r \sin \left(\frac{k\pi}{n} - \frac{\theta}{2} \right)$. Then

$$\sum^n_{k=1} (-1)^k a_k \cos \frac{\pi}{2n} = \sum (-1)^k r \left(\sin \left(\frac{(2k-1)\pi}{2n} - \frac{\theta}{2} \right) + \sin \left(\frac{(2k+1)\pi}{2n} - \frac{\theta}{2} \right) \right)$$

$$= \left(-1 + (-1)^{n+1} \right) r \sin \left(\frac{\pi}{2n} - \frac{\theta}{2} \right),$$

which leads us to

$$\sum^n_{k=1} (-1)^k a_k = \begin{cases} 2r \sin \left(\frac{\theta}{2} - \frac{\pi}{2n} \right) \left(\cos \frac{\pi}{2n} \right)^{-1}, & \text{for even } n, \\ 0, & \text{for odd } n. \end{cases}$$

Problem 3.31. Consider n distinct complex numbers $z_i \in \mathbb{C}$ such that

$$\min_{i \neq j} |z_i - z_j| \geq \max_{i \leq n} |z_i|.$$

What is the greatest possible value of n?

Solution 3.31. Let $|z_n| = \max_{i \leq n} |z_i|$; then the z_i belong to the disk of radius $R = |z_n|$ centered at the origin. Moreover, the distance between two points of complex coordinates z_i and z_j is given by $|z_i - z_j|$, which by hypothesis is greater than R. In particular, we have n points whose pairwise distances are at least R within the circle of radius R. Let O be the origin and Z_i the points. We order the points Z_i in increasing order with respect to their arguments. If the origin is among the Z_i, then we choose it to be the first point. Further, there remain at least $n - 1$ points distinct from the origin. One knows that

$$\sum_{i=2}^{n} \widehat{Z_i O Z_{i+1}} = 2\pi,$$

and thus there exists at least one angle $\widehat{Z_i O Z_{i+1}} \leq \frac{2\pi}{n-1}$. In particular, if $n \geq 8$, then $|Z_i Z_{i+1}| \leq R\sqrt{2 - 2\cos \widehat{Z_i O Z_{i+1}}} < R$, contradicting our assumptions. Thus $n \leq 7$ and the maximal value $n = 7$ is reached by the configuration of six vertices of a regular hexagon together with the origin.

Problem 3.32. The interior of a triangle can be tiled by $n \geq 9$ pentagonal convex surfaces. What is the minimal value of n such that a triangle can be tiled by n hexagonal strictly convex surfaces?

Solution 3.32. There is an obvious recurrence, showing how to go from k to $k + 3$.

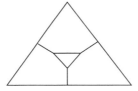

We still have to find the appropriate tiling for $n \in \{9, 10, 11\}$. If $n = 9$, then we have the following tiling of the triangle:

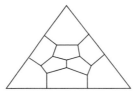

A slight deformation of the left corner pentagon shows that a concave quadrilateral can also be tiled with nine pentagons. Moreover, a tiling by n pentagons of a concave quadrilateral can be adjusted as below to a tiling of a triangle with $n + 1$ pentagons. This method furnishes a tiling with any number $n \geq 9$ of pentagons.

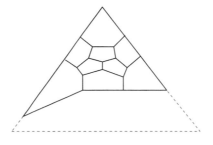

2. We will show that there are no hexagonal tilings of triangles with convex tiles. Assume the contrary. We call a tiling good if no hexagon tile has sides in common with (at least) two sides of the triangle.

Given an arbitrary tiling, we can construct a good tiling of a triangle as follows. First, the triangle can be assumed to be a right triangle by means of the following transformation. Consider the cone in $\mathbb{R}^3$ over the initial tiled triangle with a cone vertex far away, and cut it by a convenient plane so that one of the angles of the section become right angled. The trace of the coned tiling in the planar section is still a hexagonal tiling with convex polygons. Further, we double the right triangle along one of its sides and obtain a tiling that has no hexagons with common edges with both of the sides opposite to the doubled side. Applying again the same procedure yields a good tiling.

The main feature of a good tiling by convex polygons is that any hexagon having a common side with the triangle (called a boundary hexagon) has precisely one side in common, excepting the case in which the hexagon might have a common angle with the triangle (called an exceptional boundary hexagon). In the last case, we have a vertex of the hexagon and its incident sides that are common with the vertex of a triangle and its incident sides. Thus there are $\lambda \in \{0, 1, 2, 3\}$ exceptional boundary hexagons. Let a_∂ be the total number of boundary hexagons and a_{in} the hexagons that are interior, i.e., they have no sides in common with the triangle.

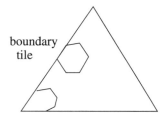

boundary
tile

exceptional
boundary tile

Construct the dual graph associated to the tiling. This is a planar graph whose vertices have degree 6. The edges going outside of the triangle will be called exterior edges of the graph, and the remaining ones interior edges. The number of vertices of this graph is $a = a_{in} + a_\partial$. Counting the sum of degrees of all vertices, we obtain $6a$ on one side, and $2e_{in} + e_o$, where e_{in} (respectively e_o) is the number of interior (respectively exterior) edges. Moreover, each exterior edge comes from a unique boundary hexagon, and two different edges come from different hexagons, excepting the exceptional hexagons, which have two outgoing edges. Thus $e_o = a_\partial + \lambda$. This implies that $e_{in} = 3a - \frac{1}{2}(a_\partial + \lambda)$.

If we erase the exterior edges of our graph, then we obtain a planar polygon partitioned into polygons (duals of the hexagons). The total number of vertices is a, the number of edges is e_{in} and the number of faces f can be estimated from above as follows. We have a_{in} vertices in the interior of a polygon with a_∂ sides. Each polygon has at least three sides (because our initial hexagons were supposed to be strictly convex). Consider the sum of angles of all polygons involved. On the one hand this yields $2\pi a_{in} + \pi(a_\delta - 2)$, and on the other hand this is at least πf. Thus $f \leq 2a_{in} + a_\partial - 2$. The Euler–Poincaré theorem requires that

$$1 = a - e_{in} + f \leq a - \left(3a - \frac{1}{2}(a_\partial + \lambda)\right) + 2a_{in} + a_\partial - 2 = \frac{\lambda}{2} - 2 - \frac{a_\partial}{2}.$$

But $\lambda \leq 3$ and $a_\partial \geq 0$ lead us to a contradiction.

Comments 59 *The same proof shows that there does not exist a tiling of the quadrilateral by strictly convex hexagons.*

Problem 3.33. We say that a transformation of the plane is a congruence if it preserves the length of segments. Two subsets are congruent if there exists a congruence sending one subset onto the other. Show that the unit disk cannot be partitioned into two congruent subsets.

Solution 3.33. Let us assume that the disk D can be partitioned as $D = A \cup B$, where A and B are congruent. Let p denote the center of D, and assume that $p \in A$. Let $F : A \to B$ be a congruence.

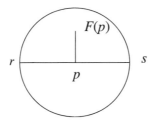

Let rs be the diameter that is perpendicular to the line $pF(p)$. For any point $x \in D$, we have $|px| \leq 1$. Thus $|F(p)F(x)| \leq 1$, for any $x \in A$, and thus $|F(p)y| \leq 1$, for any $y \in B$. Now $|F(p)r| > 1$ and $|F(p)s| > 1$, and thus $r, s \in A$. Further, $|F(r)F(s)| = |rs| = 2$ and thus $F(r)F(s)$ is a diameter of the disk. Since $|F(r)F(p)| = |F(s)F(p)| = 1$, we derive that $F(p)$ is the midpoint of the diameter $F(r)F(s)$, and thus it coincides with the center p. This contradiction proves the claim.

Comments 60 *Actually, the same argument shows that the disk cannot be partitioned into finitely many congruent sets. The problem is due to B.L. Van der Waerden. More generally, D. Puppe showed that the same holds for any convex compact set in the plane. See also:*

- H. Hadwiger and H. Debrunner: *Combinatorial Geometry in the Plane,* translated by V. Klee. With a new chapter and other additional material supplied by the translator, New York, 1964.

Problem 3.34. Prove that the unit disk cannot be partitioned into two subsets of diameter strictly smaller than 1, where the diameter of a set is the supremum distance between two of its points.

Solution 3.34. Observe first that the diameter of a planar set A coincides with the diameter of its closure $\overline{A}$ in $\mathbb{R}^2$. Assume that the disk is partitioned into two sets A and B of smaller diameter. The points of the boundary circle cannot belong to only one of the partition sets, since opposite points are at distance 1. One the other hand, the unit circle cannot be written as the union of two disjoint nontrivial closed subsets (i.e., closed intervals) because it is connected. This implies that there exists some point M of the unit circle that belongs to both $\overline{A}$ and $\overline{B}$. Consider now the opposite point M' on the circle. If $M' \in \overline{A}$, then A has diameter 1; otherwise, $M' \in \overline{B}$ and so B has diameter 1, which gives us a contradiction.

Problem 3.35. A continuous planar curve L has extremities A and B at distance $|AB| = 1$. Show that, for any natural number n there exists a chord determined by two points $C, D \in L$ that is parallel to AB and whose length $|CD|$ equals $\frac{1}{n}$.

Solution 3.35. Assume that L does not admit chords CD parallel to AB either of length a or of length b. We will show that L does not admit chords CD parallel to AB of length $a + b$. Let T_x be the planar translation along the line AB of the number of x units. We have, by hypothesis,

$$T_a L \cap L = \emptyset \text{ and } T_b L \cap L = \emptyset.$$

This implies that $T_{a+b}L \cap T_a L = \emptyset$. On the other hand, let $d(U, V)$ denote the distance between the sets U and V in the plane, given by $\inf_{P \in U, Q \in V} |PQ|$. Then obviously $d(T_{a+b}L, L) \leq d(T_a L, L) + d(T_{b+a}L, T_a L)$. This implies that $d(T_{a+b}L, L) = 0$ if $d(T_a L, L) = d(T_{b+a}L, T_a L) = 0$. Moreover, L and $T_{a+b}(L)$ are closed subsets, and thus the distance is attained, i.e., there exist $P \in L$ and $Q \in T_{a+b}(L)$ such that $|PQ| = d(L, T_{a+b}(L))$. In particular, $P = Q$ and thus $T_{a+b}L \cap L \neq \emptyset$, proving our claim.

Suppose now that there is no parallel chord of length $a = b = \frac{1}{n}$. Then we derive that there is no such chord of length $\frac{2}{n}$. Take $a = \frac{1}{n}$ and $b = \frac{2}{n}$. The previous claim shows that there is no chord of length $\frac{3}{n}$. An easy induction shows that there is no chord of length $\frac{k}{n}$ for $k \leq n$. This contradicts the fact that AB has length 1.

Comments 61 *The result is due to H. Hopf, who characterized all the possible lengths of chords arising as above.*

- H. Hopf: *Über die Sehnen ebener Kontinuen und die Schleifen geschlossener Wege,* Comment. Math. Helv. 9 (1937), 303–319.

Problem 3.36. The diameter of a set is the supremum of the distance between two of its points. Prove that any planar set of unit diameter can be partitioned into three parts of diameter no more than $\frac{\sqrt{3}}{2}$.

Solution 3.36. The main ingredient is to show that a planar set F of unit diameter is contained within a regular hexagon whose opposite sides are at distance 1.

Given a vector $v = v_1$, let us consider the thinest strip S_v parallel to v that contains F. Then the strip S_v is bounded by two parallel lines, each line containing at least one point of F, by the minimality of the strip. In particular, the width of the strip is bounded by the distance between two points of F, and so by the diameter of F.

Now let v_2 be the image of v_1 by a rotation of angle $\frac{\pi}{3}$, and let v_3 be the image of the rotation of angle $\frac{2\pi}{3}$. The intersection of the corresponding strips $H_v = S_{v_1} \cap S_{v_2} \cap S_{v_3}$ is a semiregular (i.e., all opposite edges are parallel) hexagon having all angles $\frac{2\pi}{3}$. However, in general, this hexagon is not regular.

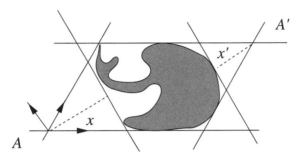

We choose two opposite vertices, say A and A', of the parallelogram $S_{v_1} \cap S_{v_2}$. For instance, we choose A such that the vectors spanning the parallelogram issued from it are positive multiples of both v_1 and v_2. We denote by x the distance from A to the strip S_{v_3} and by x' the distance from A' to the strip S_{v_3}. We observe that the hexagon H_v is regular if $x = x'$. Moreover, let us set $f(v) = x - x'$. Obviously, the function that associates to the vector v the quantity $x - x'$ is continuous. Let us compute $f(-v)$, by rotating v through π. This amounts to interchanging A with A', and thus $f(-v) = -f(v)$. A continuous function taking both positive and negative values must have a zero. Thus there exists some position of v for which $f(v) = 0$ and so H_v is regular.

Further, the regular hexagon of width 1 can be partitioned into three pieces having diameter $\frac{\sqrt{3}}{2}$, as in the figure below:

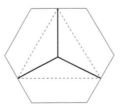

This proves the claim.

Comments 62 *K. Borsuk proved in 1933 that a planar set of diameter 1 can be partitioned into subsets of diameter $d < 1$. The main ingredient above is a lemma due to J. Pál, from 1920. B. Grünbaum proved (improving on ideas of D. Gale) that a set of unit diameter in $\mathbb{R}^3$ is contained in a regular octahedron of width 1, with three of its vertices (one vertex on each diagonal) cut off by planes orthogonal to the diagonals,*

*at distance $\frac{1}{2}$ from the center. Using a suitable dissection of this polyhedron, one finds
that the set F can be partitioned into four pieces having diameter not bigger than*

$$\sqrt{\frac{6129030 - 937419\sqrt{3}}{1518\sqrt{2}}} \approx 0.9887.$$

*Gale's conjecture from 1953, that one can always get a partition into four pieces of
diameter no bigger than $\sqrt{\frac{3+\sqrt{3}}{6}}$, is still open.*

- D. Gale: *On inscribing n-dimensional sets in a regular n-simplex*, Proc. Amer.
  Math. Soc. 4 (1953), 222–225.
- B. Grünbaum: *A simple proof of Borsuk's conjecture in three dimensions*, Math.
  Proc. Cambridge Philos. Soc. 53 (1957), 776–778.

Problem 3.37.   1. Prove that a finite set of n points in $\mathbb{R}^3$ of unit diameter, can be
covered by a cube of side length $1 - \frac{2}{3n(n-1)}$.
  2. Prove that any planar set of n points having unit diameter can be partitioned into
three parts of diameter less than $\frac{\sqrt{3}}{2} \cos \frac{2\pi}{3n(n-1)}$.

Solution 3.37. 1. A direction will mean below a line without specified orientation. The
angle between two directions is the smallest among the four angles they define, and
thus it is always within the range $\left[0, \frac{\pi}{2}\right]$. Consider the set of n points $\{x_1, x_2, \ldots, x_n\}$,
which determine at most $N = \frac{n(n-1)}{2}$ distinct directions $x_i x_j$, which we denote
by $l_1, l_2, \ldots, l_N$. We want to prove first that there exist three mutually orthogonal
directions y_1, y_2, y_3, so that

$$\min_{1 \leq i \leq 3, 1 \leq s \leq N} \angle(y_i, l_s) \geq \arccos\left(1 - \frac{1}{3N}\right).$$

Consider an orthogonal frame in $\mathbb{R}^3$ given by three mutually orthogonal vectors
e_1, e_2, e_3, and choose the rotation R_1 and R_2 of the space with the property that
$R_1 e_1 = e_2, R_1 e_3 = e_3, R_2 e_1 = e_3, R_2 e_2 = e_2$. Using these rotations, we have

$$\min_{1 \leq i \leq 3, 1 \leq s \leq N} \angle(y_i, l_s) = \min_{1 \leq s \leq N} \min(\angle(y_1, l_s), \angle(y_1, R_1 l_s), \angle(y_1, R_2 l_s)).$$

Consider the set of $3N$ directions $D = \{l_j, R_1 l_j, R_2 l_j, 1 \leq j \leq N\}$. We will prove
the following estimate from below:

$$\max_y \min_{l \in D} \angle(y, l) \geq \arccos\left(1 - \frac{1}{3N}\right) = \theta,$$

for an arbitrary set D of $3N$ directions. Assume the contrary. The set of directions
y making an angle smaller than θ with a given direction l form a symmetric cone
$C_{l,\theta}$ of angle 2θ with axis l. If there is no direction y satisfying the inequality above,
then the union of all cones $C_{l,\theta}$, over $l \in D$ will cover the set of all directions in
space. Each symmetric cone $C_{l,\theta}$ intersects the unit sphere along a symmetric pair of

spherical caps of radius θ. In particular, these spherical caps cover the surface of the whole sphere, since a point not covered yields a direction not belonging to the union. Furthermore, this implies that the total area of the spherical caps is (strictly) greater than the area of the sphere. Now, the area of each spherical cap is $2\pi(1 - \cos\theta)$, and thus the total area covered by the cones is $12\pi N(1 - \cos\theta) = 4\pi$. But 4π is the total area of the sphere, which contradicts the claim. Observe that the inequality must be strict, since $N \geq 1$ and any covering of the sphere by nontrivial spherical caps will have some overlaps.

Consider now three mutually orthogonal directions y_j satisfying the claim above. Let further $\xi_j \subset y_j$ be the image of the given set of points under orthogonal projection onto y_j. The distance between the images of x_i and x_k on y_j is $|x_i x_k| \cos \angle(y_j, x_i x_k) \leq \sin\theta$. Thus each subset ξ_j is contained in some interval of length less than $\cos\theta$, and thus the set of points is contained in the strip of width $\sin\theta$, orthogonal to y_j. The intersection of the three orthogonal strips associated to the direction y_j is a cube of side length $\cos\theta$, and the claim follows.

2. The second part follows using the same idea. This time, we consider three directions y_1, y_2, y_3 in the plane such that y_2 is obtained by a counterclockwise rotation R_1 of angle $\frac{\pi}{3}$ from y_1, and y_3 by a counterclockwise rotation R_2 of angle $\frac{\pi}{3}$ from y_2. We claim that for any N directions in the plane, we have

$$\min_{1\leq i\leq 3, 1\leq s\leq N} \angle(y_i, l_s) \geq \frac{\pi}{6N} = \alpha.$$

The same trick as above shows that the $3N$ symmetric angle sectors of angle 2α cover the set of all directions and thus the unit circle. Now, the total length of arcs covered by these symmetric cones is $12N\alpha = 2\pi$, and the claim follows.

Therefore, the set of points is contained within the intersection of three strips of width $w = \cos\left(\frac{2\pi}{3n(n-1)}\right)$, which are orthogonal, respectively, to the directions y_j. The intersection is a semiregular hexagon H. Since all three strips have the same width, we find that there are only two different side lengths, as in the figure below:

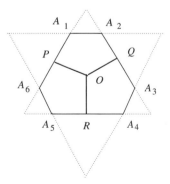

This means that H is the intersection of two equilateral triangles of the same incenter O. Let $A_1, \ldots, A_6$ be the vertices of H. We have then $|A_1 A_2| = |A_3 A_4| = |A_5 A_6| = a$ and $|A_2 A_3| = |A_4 A_5| = |A_6 A_1| = b$. Moreover, by computing the width of H, we find that the lengths a, b are constrained to satisfy $a + b = \frac{2}{\sqrt{3}} w$. Assume from now on that $a \leq b$. We divide the hexagon H by means of three segments OP, OQ, OR orthogonal to the three longest sides of the hexagon. The diameter of the pentagon OPA_1A_2Q is one of its diagonals, and one computes easily $|PQ| = a + \frac{b}{2}$, $|OA_1| = \frac{\sqrt{3b^2 + (2a+b)^2}}{2\sqrt{3}}$, and $|PA_2| \leq |PQ|$ since $\angle(PA_2Q) \geq \angle(PQA_2) = \frac{\pi}{3}$. Using the relation between a and b, we obtain that $|PQ| \leq \frac{3}{4}(a + b) \leq \frac{\sqrt{3}}{2} w$, with equality only when $a = b$ and the hexagon is regular. On the other hand, $|OA_1| \leq \frac{2w}{3}$, with equality when $a = 0$. This implies that the diameter of each piece is at most $\frac{\sqrt{3}}{2} w$, where w has the value found above.

Comments 63 *This version of Borsuk's problem for finite sets and its higher-dimensional generalization for covering by cubes was given in:*

- L. Funar: *A minimax problem in geometry* (in Romanian), Proc. 1986 Nat. Conf. Geometry and Topology - Tîrgoviste, Univ. Bucharest, 91–96, 1988.

Problem 3.38. Prove that any convex body in $\mathbb{R}^n$ of unit diameter having a smooth boundary can be partitioned into $n + 1$ parts of diameter $d < 1$.

Solution 3.38. The first step is to observe that the claim holds for the unit ball B^n. We consider the regular spherical simplex obtained by projecting the regular simplex onto the sphere from its incenter. The $n + 1$ faces of the spherical simplex form a dissection into pieces of diameter smaller than 1, since there are no opposite points lying in the same face. Otherwise, one can compute explicitly the diameter of each piece.

Consider the convex body F with smooth boundary $W \subset \mathbb{R}^n$. The only point where the smoothness is used is when we assume that there exists a unique tangent hyperplane $T_p W$ at each point p of W. Moreover, there exists a unique unit vector v_p in $\mathbb{R}^n$, based at p, that is orthogonal to the tangent hyperplane $T_p W$ and points toward the half-space that does not contain F. This permits us to define a smooth map $G : W \to S^{n-1}$ to the unit sphere centered at the origin O, usually called the Gauss map in differential geometry. We associate to the point $p \in W$ the unique point $G(p) \in S^{n-1}$ for which the vectors $OG(p)$ and v_p are parallel, and so the tangent space $T_p W$ is parallel to the tangent hyperplane $T_{G(p)} S^{n-1}$ at the point $G(p)$ on the sphere. Since F is convex and W is smooth, the map G is a smooth diffeomorphism.

Consider now the preimage by G of a partition of the unit sphere into pieces of diameter smaller than 1. This yields a partition of W into $n + 1$ pieces, which we denote by $A_1, A_2, \ldots, A_{n+1}$. We claim that each piece A_j has diameter smaller than 1. Assume the contrary. We can replace the A_j by their closures, without modifying their diameters. Now there should exist two points $x, y \in A_j$ at distance 1, which is the diameter of F. Let $T_x W$ and $T_y W$ be the tangent planes at W at x, y respectively. Then $T_x W$ and $T_y W$ are orthogonal to the segment xy. Otherwise, a small perturbation $x' \in W$ of x in some tangent direction from $T_x W$, making an obtuse angle with xy, will increase the length of $|x'y|$, but the diameter of W is no larger than 1.

Therefore $G(x)$ and $G(y)$ are opposite points on the unit sphere, and moreover, they belong to the same set of the partition, but this is impossible, since the pieces of the partition are of diameter strictly smaller than 1.

Comments 64 *The question whether any set F in $\mathbb{R}^n$ can be partitioned into $n + 1$ pieces of strictly smaller diameter is known as Borsuk's problem. Since the diameter of a set equals the diameter of its convex hull, it suffices to consider the problem the convex sets. The result in the problem above was obtained by H. Hadwiger in 1945. The case left open is that in which the convex body has corners.*

There was no significant progress in the subject until 1993, when J. Kahn and G. Kallai gave an unexpected counterexample by showing that the minimal number $f(n)$ of pieces needed for a dissection into parts of strictly smaller diameter satisfies $f(n) \geq (1.1)^{\sqrt{n}}$ for large n and that $f(n) > n + 1$ for $n \geq 1825$. The subsets that they considered are finite sets of points. This bound was subsequently lowered by N. Alon and then, by A. Hinrichs, and C. Richter to $n = 298$. A thorough treatment of the subject is in the recent book of Boltyanski, Martini, and Soltan. However, the 4-dimensional case is still open.

- V. Boltyanski, H. Martini, and P.S. Soltan: *Excursions into Combinatorial Geometry,* Springer, Universitext, 1997.
- H. Hadwiger: *Überdeckung einer Menge durch Mengen kleineren Durchmesser,* Commentarii Math. Helvet. 18 (1945/1946), 73–75, 19 (1946/1947), 71–73.
- A. Nilli: *On Borsuk's problem,* Jerusalem combinatorics '93, 209–210, Contemp. Math., 178, Amer. Math. Soc., Providence, RI, 1994.
- J. Kahn and G. Kalai: *A counterexample to Borsuk's conjecture,* Bull. Amer. Math. Soc. 29 (1993), 1, 60–62.

Problem 3.39. Let D be a convex body in $\mathbb{R}^3$ and let $\sigma(D) = \sup_\pi \text{area}(\pi \cap D)$, where the supremum is taken over all positions of the variable plane π. Prove that D can be divided into two parts D_1 and D_2 such that $\sigma(D_i) < \sigma(D)$.

Solution 3.39. Choose a point in the interior of D and a sequence of chords $l_1, l_2, \ldots,$ $l_n, \ldots$ of D passing through this point, the chords being dense in D. Choose also a sequence ε of positive numbers decreasing rapidly to zero. Let C_n be the set of points at distance less than ε_n from l_n. By choosing ε small enough, we can ensure that $\sigma(C_n) < 2^{-n}\sigma(D)$. Set $D_1 = \cup_{n=1}^\infty C_n$ and $D_2 = D - D_1$. Then $\sigma(D_1) < \sum_{n=1}^\infty \sigma(C_n) < \sigma(D)$. It is also clear that for any plane π, $\text{area}(\pi \cap D_2) < \sigma(D)$. But D_2 is a compact and $\text{area}(\pi \cap D_2)$ is a continuous function on the plane π. The set of planes that intersect D can be parameterized by a point in D (which will be considered the origin) times the projective space $\mathbb{R}P^2$, and thus by a compact set. Thus $\text{area}(\pi \cap D_2)$ achieves its minimum, which will therefore be strictly smaller than $\sigma(D)$.

Comments 65 *The problem was proposed by L. Funar and the solution above was given by C.A. Rogers. A more interesting question is to consider dissections into convex subsets. The additional convexity assumption implies that the boundary structure of* *each piece (at the interior of the body) should be polyhedral. In particular, the minimal number of convex pieces with smaller σ, which the unit ball requires, is 3.*

If the body D is strictly convex (meaning that the open segment determined by two of its boundary points is contained in the interior of D) or D is a polyhedron without parallel edges, then we can show that four pieces suffice. By a recent result of M. Meyer, two plane sections of maximal area cannot be disjoint. Fix a maximal-area cross-section F and let F(ε) be the set of points of D, at distance less than ε from F. The complement of F(ε) has two components, A^+ and A^-, which are of strictly smaller σ, because there is no maximal-area plane section contained completely in D − F(ε). Further, F(ε) can be subdivided, by a hyperplane orthogonal to F, into two pieces of smaller σ if ε is small enough. However, we don't know whether three pieces suffice for an arbitrary convex body.

A related problem is whether we have $\sigma(D) \leq \frac{1}{4}\text{area}(\partial D)$, where ∂D denotes the boundary of D. This would imply that the number of maximal area planar sections with disjoint interiors is at most 4, with equality for the regular simplex.

- L. Funar: *Problem E 3094,* Amer. Math. Monthly 92 (1985), 427.
- M. Meyer: *Two maximal volume hyperplane sections of a convex body generally intersect,* Period. Math. Hungar. 36 (1998), 191–197.

Problem 3.40. If we have k vectors $v_1, v_2, \ldots, v_k$ in $\mathbb{R}^n$ and $k \leq n + 1$, then there exist two vectors making an angle θ with $\cos\theta \geq -\frac{1}{k-1}$. Equality holds only when the endpoints of the vectors form a regular $(k − 1)$-simplex.

Solution 3.40. Assume that the vectors v_j are unit vectors. Then

$$\sum_{1\leq i<j\leq k} |v_i - v_j|^2 \leq \left(\sum_{1\leq i<j\leq k} |v_i - v_j|^2\right) + \left|\sum_{i=1}^k v_i\right|^2 = k\sum_{i=1}^k |v_i|^2 = k^2.$$

Thus there exists at least one pair i, j for which $|v_i - v_j| \leq \frac{2k}{k-1}$. This is equivalent to the fact that the angle θ between v_i and v_j satisfies the claimed inequality.

Problem 3.41.   1. Consider a finite family of bounded closed convex sets in the plane such that any three members of the family have nonempty intersection. Prove that the intersection of all members of the family is nonempty.

2. A set of unit diameter in $\mathbb{R}^2$ can be covered by a ball of radius $\sqrt{\frac{1}{3}}$.

Solution 3.41. 1. Assume that the convex sets are $A_1, A_2, \ldots, A_k$ and $k \geq 4$. We claim that if each $k − 1$ sets among them have a common point, then all of them should have a common point. The result of the problem follows by using recurrently the previous claim.

Before proceeding, we observe that given a set of $k \geq 4$ points in the plane, we can find two subsets whose convex hulls intersect nontrivially. Here the convex hull of a set of points is the smallest convex polygon containing these points (and thus all segments pairwise connecting them). In fact, if we choose four points among them, then their convex hull is either a triangle containing the fourth point in its interior, or else a quadrilateral having two diagonals crossing each other.

Let p_i be a point belonging to $\cap_{j \neq i, j \leq k} A_i$ and possibly not to A_i. There exists therefore a partition of the set $\{p_1, p_2, \ldots, p_k\} = P \cup Q$, where $P = \{p_{i_1}, p_{i_2}, \ldots, p_{i_m}\}$ and $Q = \{p_{j_1}, p_{j_2}, \ldots, p_{j_{k-m}}\}$, such that the convex hull of the subset P intersects in at least one point the convex hull of the subset Q. Recall that each point of P belongs to $\cap_{i \notin \{i_1, i_2, \ldots, i_m\}} A_i$, which is convex as the intersection of convex sets. Thus the convex hull of P is equally contained within $\cap_{i \notin \{i_1, i_2, \ldots, i_m\}} A_i$. The same argument shows that the convex hull of Q is contained in $\cap_{j \notin \{j_1, j_2, \ldots, j_{k-m}\}} A_j$. Since the two convex hulls have at least one common point, this therefore belongs to $\cap_{i=1}^{k} A_i$, as claimed.

2. It suffices to see that any finite subset is contained in a disk of radius $\sqrt{\frac{1}{3}}$, which means that the intersection of balls centered at that points of radius $\sqrt{\frac{1}{3}}$ is nonempty. By the previous claim, it suffices to consider the case of three points. If the triangle made by these points is obtuse, then we can take the disk whose diameter is the largest side of the triangle. If not, then all angles of the triangle are acute and there exists at least one angle, say $A \geq \frac{\pi}{3}$. The circumradius of the triangle is $R = \frac{a}{2 \sin A} \leq \sqrt{\frac{1}{3}}$.

Comments 66 *The first part of the problem is the particular case of the celebrated Helly's theorem, proved in 1923: if a family of closed convex sets in $\mathbb{R}^n$ has the property that any $n + 1$ among them have a common point, then the intersection of all sets is nonempty.*

The second claim was generalized by Jung in 1901, who proved that any set of diameter 1 in $\mathbb{R}^n$ admits an unique n-ball of minimal radius covering it, and the minimal radius is bounded by $\sqrt{\frac{n}{2(n+1)}}$. The original proof of Jung's theorem was quite complicated, but there exists an elementary proof given later by Blumenthal and Wahlin.

It is worthy of notice that any subset of $\mathbb{R}^n$ of diameter 1 can be covered also by a simplex of diameter $\frac{\sqrt{n(n+1)}}{2}$, according to Gale.

- L.M. Blumenthal, G.E. Wahlin: *On the spherical surface of smallest radius enclosing a bounded subset of n-dimensional Euclidean space*, Bull. Amer. Math. Soc. 47 (1941), 771–777.
- L. Danzer, B. Grünbaum, and V. Klee: *Helly's theorem and its relatives*, Convexity, Proc. Symposia Pure Math. vol. 7, AMS, 1963, 101–180.
- D. Gale: *On inscribing n-dimensional sets in a regular n-simplex*, Proc. Amer. Math. Soc. 4 (1953), 222–225.
- H.W.E. Jung: *Über die kleinste Kugel, die eine räumliche Figur einschliesst*, J. Reine Angew. Math. 123 (1901), 241–257.

Problem 3.42. Let T be a right isosceles triangle. Find the disk D such that the difference between the areas of $T \cup D$ and $T \cap D$ is minimal.

Solution 3.42. Let us show that the circle C bounding D has the following two properties, valid for an arbitrary triangle T:

1. The circle C is split into three arcs outside T and three arcs inside T. Then the total length of the outside arcs equals the total length of the inside arcs.
2. The three chords determined by the edges of T have their lengths proportional to the respective side lengths.

1. Assume that C has fixed center. When the radius of C increases by ε, area$(T-D)$ decreases as $l_{in}\varepsilon^2$, where l_{in} is the length of the inside arcs. On the other hand, area$(D-T)$ decreases as $l_{ou}\varepsilon^2$, where l_{ou} is the length of the outside arcs. Thus we have a local minimum only if $(l_{in} - l_{ou})\varepsilon^2$ does not increase and thus $l_{in} = l_{ou}$.

2. Assume now that C has fixed radius. Since area$(T - D) -$ area$(D - T) =$ area$(T) -$ area(D), it is sufficient to minimize area$(D - T)$. Consider that the center of D is slightly translated by $\varepsilon\overline{u}$, where $\overline{u}$ is a unit vector.

Then we can compute the change of area$(S - T)$ by summing the contribution of each chord sector. The contribution of $A_1 A_2$ is $\varepsilon\overline{u} \times (\overline{A_1 A_2})$ and thus the total variation is

$$\varepsilon\overline{u} \times (\overline{A_1 A_2} + \overline{B_1 B_2} + \overline{C_1 C_2}).$$

The variation must be zero in all directions u and so we obtain the condition

$$\overline{A_1 A_2} + \overline{B_1 B_2} + \overline{C_1 C_2} = 0.$$

This implies that the chords are proportional to the side lengths.

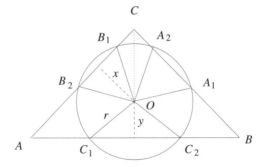

If T is right isosceles triangle of unit side length, then the center O of the circle C belongs to the altitude, by symmetry considerations. Denote by x the distance from the center O to the edges CB and CA, by y the distance from O to the edge AB. The angles determined by the outside chords are $2\alpha, 2\beta, 2\beta$ respectively. Since O belongs to the altitude, we have $y + \sqrt{2}x = \frac{1}{2}$. Moreover, the first claim above implies that $\alpha + 2\beta = \frac{\pi}{2}$ and thus $\cos\alpha = \sin 2\beta$. This translates into $\frac{y}{r} = \frac{2x}{r^2}\sqrt{r^2 - x^2}$ and thus $y^2 r^2 = 4x^2 r^2 - 4x^4$. The second statement above implies that $\sqrt{r^2 - y^2}/\sqrt{2} = \sqrt{r^2 - x^2}$, and thus introducing above, we obtain $2r^2 = 4x - 1$. In particular, x satisfies the equation

$$16x^4 - 16x^3 - 12x^2 + 8x - 1 = (4x^2 + 2x - 1)(4x^2 - 6x + 1) = 0.$$

The only convenient root is $x = \frac{\sqrt{5}-1}{4}$, since among the remaining ones, one is negative, another one is larger than 1, and the last one would lead to a negative value for r. Thus

$$r = \sqrt{\frac{1}{2}(\sqrt{5}-2)}, \quad y = \frac{1+\sqrt{2}-\sqrt{5}}{2\sqrt{2}},$$

and the circle is determined by these parameters.

Problem 3.43. Let r be the radius of the incircle of an arbitrary triangle lying in the closed unit square. Prove that $r \leq \frac{\sqrt{5}-1}{4}$.

Solution 3.43. It is clear that the inradius of a triangle contained in a larger triangle is not bigger than the inradius of the bigger triangle. Thus it suffices to consider only triangles whose vertices lie on the boundary of the square. Further, at least one vertex of the triangle can be assumed to be in a corner of the triangle. If two vertices of the triangle belong to the same edge, then we can assume that these vertices are the endpoints of that edge, as above. Thus it suffices to consider the triangle OAB whose vertices are $O = (0,0)$, $A = (a,1)$, and $B = (1,b)$ with $0 \leq a, b \leq 1$. The inradius of a triangle of area S and semiperimeter p is given by $r = \frac{S}{p}$. Therefore,

$$r = \frac{1-ab}{\sqrt{1+a^2} + \sqrt{1+b^2} + \sqrt{(1-a)^2 + (1-b)^2}}.$$

The inequality to be proved is then equivalent to

$$F(a,b) = \sqrt{1+a^2} + \sqrt{1+b^2} + \sqrt{(1-a)^2 + (1-b)^2} - (\sqrt{5}+1)(1-ab) \geq 0$$

for $0 \leq a, b \leq 1$. We have $F(a,0) \geq 0$. Further, $\frac{\partial^2 F}{\partial^2 a} > 0$ by a simple computation. Let $\varepsilon^2 = 1 - \frac{6\sqrt{3}}{(1+\sqrt{5})^2}$.

1. If $1 \leq b \leq \frac{1}{4}$, set $f(x) = F(x,b)$. Then we observed above that $f'(x)$ is increasing, and since $f'(0) = (1+\sqrt{5})y - \frac{1}{y^2-2y+2} > 0$, then $f'(x) > 0$ for any x, and thus $f(x)$ is increasing. In particular, $F(a,b) \geq F(0,b) \geq 0$, provided that $\frac{1}{4} \leq b \leq 1$.

2. If $0 \leq a \leq \frac{1}{4}$ and $0 \leq y \leq \varepsilon$, we have

$$f'(x) = (1+\sqrt{5})b + \frac{x}{\sqrt{1+x^2}} - \frac{1-x}{\sqrt{(1-x)^2+(1-b)^2}}$$

$$\leq (1+\sqrt{5})\varepsilon + \frac{x}{\sqrt{1+x^2}} - \frac{1-x}{\sqrt{x^2-2x+2}}.$$

Thus $f'(\frac{1}{4}) < 0$. As f' is increasing, we have $f'(x) < 0$, for $x \in [0, \frac{1}{4}]$, and so f is decreasing on $[0, \frac{1}{4}]$. But $f(\frac{1}{4}) = F(\frac{1}{4}, b) = F(b, \frac{1}{4}) \geq 0$, by the previous case. Thus $F(a,b) \geq 0$, for $0 \leq a \leq \frac{1}{4}$ and $0 \leq y \leq \varepsilon$.

3. Since F is symmetric, it suffices to check now the case $\varepsilon \leq a \leq \frac{1}{4}$ and $\varepsilon \leq b \leq \frac{1}{4}$. Actually, we will use only the fact that $ab \geq \varepsilon^2$. This amounts to saying

that the area S of the triangle OAB is $S = \frac{1-ab}{2} \leq \frac{1-\varepsilon^2}{2}$. Any triangle satisfies the inequality $S \leq \frac{p^2}{3\sqrt{3}}$, obtained from the Heron formula for the area by the means inequality. In our case, we have

$$r = \frac{S}{p} \leq \frac{S}{\sqrt{3\sqrt{3}S}} \leq \sqrt{5} + 1.$$

This settles our claim.

Comments 67 *The solution presented here is due to Abi-Khuzam and Barbara:*

- L. Funar: *Problem 6477,* Amer. Math.Monthly 81 (1984), 588.
- F. Abi-Khuzam, R. Barbara: *A sharp inequality and the inradius conjecture,* Math. Inequalities and Appl. 4 (2001), 323–326.

Problem 3.44. Let P be a point in the interior of the tetrahedron $ABCD$, with the property that $|PA| + |PB| + |PC| + |PD|$ is minimal. Prove that $\widehat{APB} = \widehat{CPD}$ and that these angles have a common bisector.

Solution 3.44. Let $f_A(x) = |AX|$. Then $f_A : E^3 \to E^3$ is a smooth function on $E^3 - \{A\}$. Moreover, the gradient function $(\nabla f_A)(X)$ is the unit vector of direction AX, namely $\frac{AX}{|AX|}$. Consider now the function $g = f_A + f_B + f_C + f_D$ for $P \in \text{int}(ABCD)$. Then g having a minimum at P implies that

$$(\nabla g)(P) = \sum (\nabla f_A)(P) + \sum (\nabla f_B)(P) + \sum (\nabla f_C)(P) + \sum (\nabla f_D)(P) = 0.$$

If a, b, c, d are the unit vectors of directions AP, BP, CP, DP, then the previous conditions reads $a + b + c + d = 0$, or alternatively, $-(a+b) = c+d$. Observe now that $(a+b)$ is the direction of the bisector of the angle $\widehat{APB}$, while $c+d$ is the direction of the bisector of $\widehat{CPD}$. Thus these two angles have a common bisector. On the other hand, we have $\langle a + b, a + b \rangle = \langle c + d, c + d \rangle$, and because $|a| = |b| = |c| = |d|$, by simplifying the terms, we obtain that $\langle a, b \rangle = \langle c, d \rangle$. This is equivalent to saying that $\widehat{APB} = \widehat{CPD}$.

Problem 3.45. Let $OA_1, \ldots, OA_n$ be n linearly independent vectors of lengths $a_1, \ldots, a_n$. We construct the parallelepiped H having these vectors as sides. Then consider the n altitudes in H as a new set of vectors and further, construct the parallelepiped E associated with the altitudes. If h is the volume of H and e the volume of E, then prove that

$$he = (a_1 \ldots a_n)^2.$$

Solution 3.45. Let $v_i = OA_i$ and w_i be the altitude vectors. We have the relations

$$\langle w_j, v_i \rangle = \delta_{ij} \|v_i\|^2,$$

where δ_{ij} is Kronecker's symbol and $\langle , \rangle$ is the scalar product. The linear functionals $f_i(v_j) = \delta_{ij}\|v_i\|^2$ are linearly independent, and thus the dual vectors w_j are linearly independent. Thus E is well defined.

Moreover, let A be the matrix consisting of the components of the v_i's and let B be the matrix consisting of the components of the w_j's. It follows that

$$AB^\top = \operatorname{diag}\left(\|v_1\|^2, \ldots, \|v_n\|^2\right),$$

where $\operatorname{diag}(c_1, \ldots, c_n)$ denotes the diagonal matrix with given entries. Therefore

$$\operatorname{vol}(E)\operatorname{vol}(H) = |\det\ A| \cdot |\det\ B^\top| = (a_1 \ldots a_n)^2.$$

Problem 3.46. Let F be a symmetric convex body in $\mathbb{R}^3$ and let $A_{F,\lambda}$ denote the family of all sets homothetic to F in the ratio λ that have only boundary points in common with F. Set $h_F(\lambda)$ for the greatest integer k such that $A_{F,\lambda}$ contains k sets with pairwise disjoint interiors. Prove that

$$h_F(\lambda) \leq \frac{(1 + 2\lambda)^3 - 1}{\lambda^3}.$$

Solution 3.46. Set $B_\lambda = \cup_{H \in A_{F,\lambda}} H$. We prove first that

$$B_\lambda \subset (1 + 2\lambda)F,$$

where αF denotes the homothety of F in ratio α. If v is a point of the boundary ∂F, let v' be given by $|ov'| = (1 + \lambda)|ov|$, where o is the symmetry center of F. Set F_v for the set in $A_{F,\lambda}$ having center at v'. Let x be a point on the boundary of F_v, $ox \cap \partial F = \{a\}$, $ov \cap \partial F_v = \{q\}$, and let vx'' be parallel to qx, with $x'' \in \partial F$. Then

$$\widehat{qv'x} = \widehat{v'xo} + \widehat{v'ox} \geq \widehat{v'ox},$$

so that

$$\widehat{vox} \leq \widehat{vox''} = \widehat{qv'x}.$$

Since F is convex, the intersection of the two segments $|oa| \cap |vx''|$ is nonempty and contains at least one point, say b. Then $|vx''| \subset F$, $b \in F$, $b \in |ox|$. Since

$$\frac{\|oa\|}{\|ox\|} \geq \frac{\|ob\|}{\|ox\|} = \frac{\|ov\|}{\|oq\|} = \frac{1}{(1 + 2\lambda)},$$

the point x belongs to $(1 + 2\lambda)F$.

Now assume that we have k subsets $F_i \in A_{F,\lambda}$ that have pairwise disjoint interiors. Then their union is contained in B_λ, and all of them are disjoint from F. This means that their union is contained in $(1 + 2\lambda)F - F$. Their total volume cannot therefore exceed $\operatorname{vol}((1 + 2\lambda)F - \operatorname{vol}(F)$, and thus their number is bounded by

$$\frac{1}{\operatorname{vol}(F)}\left(\operatorname{vol}(1 + 2\lambda)F - \operatorname{vol}(F)\right) = \frac{(1 + 2\lambda)^3 - 1}{\lambda^3}.$$

Comments 68 *Hadwiger considered the problem of finding estimates for $h_F(1)$ (which is called the Hadwiger number of F) and proved in 1957 that*

$$n^2 + n \leq h_F(1) \leq 3^n - 1$$

for a convex body F in $\mathbb{R}^n$. Grünbaum conjectured in 1961 (and proved for n = 2) that for any r satisfying these inequalities there exists a convex body F such that $h_F(1) = r$.

More generally, the proof used above shows that

$$h_F(\lambda) \leq \frac{(1 + 2\lambda)^n - 1}{\lambda^n}.$$

We have equality above only if $\frac{1}{\lambda} \in Z_+$ and F is a parallelohedral body. The result is due to V. Boju and L. Funar.

There are no good estimates from below except in dimension 2. In fact, if n = 2, the function $\lambda h_F(\lambda)$ approaches an interesting invariant of the oval F, called the intrinsic perimeter. Define the norm $\| * \|_F$ by means of

$$\|xy\|_F = \frac{\|xy\|}{\|oz\|},$$

where z is a point of the boundary ∂F such that xy and oz are parallel. This is called the Minkowski norm (or metric) defined by the oval F. The intrinsic perimeter $p(F)$ of ∂F is the limit of the perimeters of polygons inscribed in F that approach ∂F. For instance, the intrinsic perimeter of a circle is 2π and that of a square is 8. Golab in 1932 and Reshetnyak in 1953 showed that the intrinisc perimeter satisfies the inequalities

$$6 \leq p(F) \leq 8$$

with equality in the left side only when F is a hexagon, and on the right side only when F is a rectangle.

Then it was proved by V. Boju and L. Funar that the intrinisc perimeter is related to Hadwiger numbers by means of the formula

$$p(F) = 2 \lim_{\lambda \to 0} \lambda h_F(\lambda),$$

and moreover, we have

$$3 + \frac{3}{\lambda} \leq h_F(\lambda) \leq 4 + \frac{4}{\lambda},$$

with equality on the left side only if $\frac{1}{\lambda} \in Z_+$ and F is an affine regular hexagon.

More about packing and covering invariants of convex bodies can be found in the recent book of Böröckzy, and a complete survey on the Minkowski geometry in the book of Thompson.

- V. Boju and L. Funar: *Generalized Hadwiger numbers for symmetric ovals,* Proc. Amer. Math. Soc. 119 (1993), 931–934.
- K. Böröczky Jr.: *Finite packing and covering,* Cambridge Tracts in Mathematics, 154, Cambridge University Press, Cambridge, 2004.
- H. Hadwiger: *Über Treffanzahlen be. translationsgleichen Eikörpern,* Archiv Math. 8 (1957), 212–213.

• A.C. Thompson: *Minkowski geometry,* Encyclopedia of Mathematics and Its Applications, 63. Cambridge University Press, Cambridge, 1996.

Problem 3.47. Let Δ denote the square of equations $|x_i| \leq 1$, $i = 1, 2$, in the plane, and let $A = (a_1|a_2)$ be an arbitrary nonsingular 2×2 matrix partitioned into two columns. We identify each column with a vector in $\mathbb{R}^2$. Prove that the following inequality holds:

$$\min_A \max_{x\in\Delta} \left| \frac{\langle a_1, x\rangle \langle a_2, x\rangle}{\det A} \right| = \frac{1}{2},$$

where $\langle x, y\rangle = x_1 x_2 + y_1 y_2$ is the usual scalar product.

Solution 3.47. Let $f(A, x)$ be the function under minimax. Taking $A_0 = \begin{pmatrix} 1 & 1 \\ -1 & 1 \end{pmatrix}$, we obtain

$$\max_{x\in\Delta} |f(A_0, x)| = \frac{1}{2}.$$

Thus it suffices to show that for general A, we have

$$\max_{x\in\Delta} |f(A, x)| \geq \frac{1}{2}.$$

This is equivalent to showing that

$$|\det A| \leq 1, \quad \text{provided that } \max_{x\in\Delta} |\langle a_1, x\rangle \langle a_2, x\rangle| = \frac{1}{2}.$$

Assume that the contrary holds, and so $|\det A| > 1$. The set of points in the plane with coordinates $(\langle a_1, x\rangle, \langle a_2, x\rangle) \in \mathbb{R}^2$, for x ranging in the square Δ, is a parallelogram π centered at the origin O and having area $4|\det A| > 4$.

According to our claim, the parallelogram π is contained in the planar region $Z = \{(u, v) \in \mathbb{R}^2; |uv| \leq \frac{1}{2}\}$. Let us consider the point $P = (0, s)$, where $s > 0$, is a boundary point of π, and let $L \subset \mathbb{R}^2$ be the lattice generated by the vectors $(1/s, s/2)$ and $(-1/s, s/2)$. The lattice has area $\frac{1}{2}$. Then, according to Minkowski's theorem, the interior of π contains at least one point of the lattice L, different from the origin. Let this point be given by

$$m(1/s, s/2) + n(-1/s, s/2) = ((m - n)/s, (m + n)s/2), \text{ where } m, n \in \mathbb{Z}.$$

We can also suppose $\gcd(n, m) = 1$. Since the interior of π is contained in Z, it follows that $|m^2 - n^2| < 1$ and thus $m^2 = n^2$. The only possibilities are $(0, \pm s)$ and $(\pm 2/s, 0)$. The points $(0, \pm s)$ are on the boundary of π and they are not convenient. If $Q = (2/s, 0) \in \text{int}(\pi)$, then the midpoint S of the segment $|PQ|$ should also belong to the interior of π. However, $S = (1/s, s/2)$ belongs to the boundary curve of equation $|uv| = 1/2$ of the region Z containing π, so it cannot lie within the interior of π. This contradiction proves our claim.

Problem 3.48. We denote by $\delta(r)$ the minimal distance between a lattice point and the circle $C(O, r)$ of radius r centered at the origin O of the coordinate system in the plane. Prove that

$$\lim_{r\to\infty} \delta(r) = 0.$$

Solution 3.48. Let G be a lattice point on the axis Ox such that $|OG| = [r]$. Draw a perpendicular line at G to Ox that intersects the circle $C(O, r)$ at the point C situated between the lattice points A and B, in the upper half-plane. We choose B to be the point that lies in the interior of the disk $D(O, r)$ of radius r. Notice that $|AB| = 1$, because these are neighboring lattice points on the same axis line.

Let us now draw $BD \perp AB$, where D is again a lattice point, $|BD| = 1$, and D is situated outside the disk $D(O, r)$.

Since the problem asks for the value at the limit of $\delta(r)$ when $r \to \infty$, we can assume that $r > 4$.

Let now consider the line CF that is tangent to the circle $C(O, r)$ and such that $F \in BD$. We show first that $|BF| \leq |BD|$.

Assume the contrary, namely that $|BF| > |BD|$. Then $\widehat{CFB} < \widehat{CDB} \leq \widehat{ADB} = \frac{\pi}{4}$. Furthermore,

$$\sin \widehat{CFB} = \sin \widehat{OCG} = \sin\left(\arcsin \frac{[r]}{r}\right) > \frac{r-1}{r} > \frac{\sqrt{2}}{2} = \sin \frac{\pi}{4}, \text{ if } r > 4.$$

Thus $\widehat{CFB} > \frac{\pi}{4}$, which is false.

Consequently, $|BF| \leq |BD| = 1$. The triangles CBF and OCG are similar, and therefore

$$\frac{|BC|}{|OG|} = \frac{|CG|}{|OG|} = \frac{\sqrt{r^2 - [r]^2}}{2} < \frac{\sqrt{r^2 - (r-1)^2}}{r} < \frac{\sqrt{2r}}{r} = \frac{\sqrt{2}}{\sqrt{r}}.$$

Now it is clear that

$$\delta(r) \leq |BC| \leq \frac{|BC|}{|BF|} = \frac{\sqrt{2}}{\sqrt{r}}, \text{ for } r > 4.$$

Thus $\lim_{r \to \infty} \delta(r) = 0$.

Problem 3.49. Consider a curve C of length l that divides the surface of the unit sphere into two parts of equal areas. Show that $l \geq 2\pi$.

Solution 3.49. Let C' be the symmetric (opposite, or antipodal) of C with respect to the center O of the sphere. If $C \cap C' = \emptyset$, then int $C \cap$ int $C' = \emptyset$ and thus the area of the sphere is strictly greater than the sum of areas of int C and int C'. By hypothesis, the latter is just the area of the sphere, which is a contradiction.

Therefore, $C \cap C'$ contains at least one point, say P. The point P', the symmetric of P with respect to the symmetry center of $C \cap C'$, should also belong to $C \cap C'$, since the later is symmetric. Finally, note that the shortest curve on the sphere that joins two opposite points is an arc of a great circle, and since C is made of two different arcs joining P and P', it follows that $l \geq 2\pi$.

Comments 69 *More generally, let K be a body with a symmetry center O and set $r = \min_{P \in \partial K} |OP|$, where ∂K denotes the boundary surface of K. If $C \subset \partial K$ is a curve that separates two regions on ∂K of the same area, then the length l of C satisfies $l \geq 2\pi r$.*

Problem 3.50. Let K be a planar closed curve of length 2π. Prove that K can be inscribed in a rectangle of area 4.

Solution 3.50. Let $b_K(\theta)$ be the width of the figure K in the direction θ, identified with a vector $\theta \in S^1$. If $R(K, \theta)$ denotes the area of the rectangle circumscribing K and having one side parallel to θ, then

$$R(K, \theta) = b_K(\theta) b_K \left(\theta + \frac{\pi}{2} \right).$$

Thus

$$\min_\theta R(K, \theta)^{\frac{1}{2}} \le \frac{1}{2\pi} \int_0^{2\pi} b_K(\theta)^{\frac{1}{2}} b_K \left(\theta + \frac{\pi}{2} \right)^{1/2} d\theta,$$

with equality holding iff $b_K(\theta) b_K \left(\theta + \frac{\pi}{2} \right)$ is constant. From the Cauchy–Schwartz inequality, one derives

$$\min_\theta R(K, \theta) \le \frac{1}{4\pi^2} \left(\int_0^{2\pi} b_K(\theta) \, d\theta \right)^2,$$

the equality holding iff K is a curve of constant width. We are able now to apply the following theorem of Cauchy, which expresses the length $L(K)$ of the curve K in terms of the integral of its width with respect to all directions:

$$L(K) = \frac{1}{2} \int_0^{2\pi} b_K(\theta) \, d\theta.$$

This implies that

$$\min_\theta R(K, \theta) \le \frac{1}{\pi^2} L(K)^2 \le 4,$$

where the second inequality uses the hypothesis $L(K) \le 2\pi$.

Comments 70 *We can consider the parallelograms with a given acute angle ϕ instead of rectangles. The same argument shows that*

$$\min_\theta R_\phi(K, \theta) \le \frac{\cos \phi}{\pi^2} L(K)^2.$$

See also:

- E. Lutwak: *On isoperimetric inequalities related to a problem of Moser*, Amer. Math. Monthly 86 (1979), 476–477.

Problem 3.51.   1. Consider a family of plane convex sets with area a, perimeter p, and diameter d. If the family covers the area A, then there exists a subfamily with pairwise disjoint interiors that covers at least area λA, where $\lambda = \frac{a}{a + pd + \pi d^2}$.
2. Assume that any two members of the family have nonempty intersection. Prove that there exists then a subfamily with pairwise disjoint interiors that covers area at least μA, where $\mu = \frac{a}{\pi d^2}$.

Solution 3.51. 1. If K is a member of the family and $K(d)$ the set of points at distance less than d from K, then it is known that area$(K(d)) = a + pd + \pi d^2$. Let $\{K_1, K_2, \ldots, K_n\}$ be a maximal subfamily with disjoint interiors. Then every member of the family intersects some K_i, and thus it is contained in some $K_i(d)$. Thus the family is contained within $\bigcup_{i=1}^n K_i(d)$. Thus the area covered by all members of the family is at most $n \cdot$ area$(K_i(d)) \le A$, while the subfamily above covers $na \ge \lambda A$.

2. The diameter of the union of all sets is at most $2d$, by the assumption. Its area is therefore at most πd^2, by the well-known isoperimetric inequality. Thus $\lambda A \le a$, and one single set will cover at least λA.

Comments 71 *It is still unknown whether the second claim holds without any additional assumption on the sets.*

Problem 3.52. Let C be a regular polygon with k sides. Prove that for every n there exists a planar set $S(n) \subset \mathbb{R}^2$ such that any subset consisting of n points of $S(n)$ can be covered by C, but $S(n)$ itself cannot be included in C.

Solution 3.52. Let r be the radius of the circle inscribed in C. For $n \in \mathbb{Z}_+$, let $S(n)$ be the circle of radius $r \sec(\pi/2kn)$.

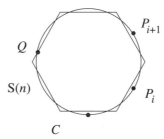

1. C cannot cover $S(n)$. In fact, there is no disk of radius greater than r that can be included in C. Here is a proof for this somewhat obvious assertion. For a fixed radius R, let $X(R)$ be the set of centers of circles of radius R included in C. Then $X(R)$ is convex and invariant of rotations centered at the center O of C, of angle $2\pi/k$. Thus $X(R)$ is either empty or contains O. Thus $R \le r$.

2. Let $P_1, \ldots, P_n \subset S(n)$ be arbitrary points of $S(n)$. We show that C can cover $P_1, \ldots, P_n$. We fix a point Q in C at a distance of $r \sec \frac{\pi}{2kn}$ from O. We place C such that their centers coincide and we rotate C around its center O such that Q runs through $S(n)$. While Q travels along the circle, some of the points P_i will be in the interior of C and the others outside.

However, we see that $P_i \notin C$ if and only if $Q \in \bigcup_{j=1}^n A_j$, where each A_j is an arc of a circle from outside C, which is of length $\frac{\pi}{kn}$. More precisely, the arc A_j is determined by the inequalities

$$2\frac{\pi j}{k} \le \widehat{P_i OQ} \le 2\frac{\pi j}{k} + \frac{\pi}{kn}.$$

Therefore $\bigcup_{j=1}^n A_j$ is a set of arcs of total length smaller than $nk\pi/kn \cdot \pi$ and thus $S - \bigcup_{j=1}^n A_j$ has length π, and thus it is nonempty. Now rotate C until Q lies inside $S - \bigcup_{j=1}^n A_j$. Then all the P_j belong to C, as claimed.

Problem 3.53. Let M be a convex polygon and let $S_1, \ldots, S_n$ be pairwise disjoint disks situated in the interior of M. Does there exist a partition $M = D_1 \cup \cdots \cup D_n$ such that D_i are convex disjoint polygons, each of which contains precisely one disk?

Solution 3.53. Let O_i, r_i be the center and the radius of the disk S_i. For a point X in the plane, let us define $h_i(X) = XO_i^2 - r_i^2$. We consider

$$D_k = \{x \in M; h_k(X) \le h_i(X), \text{ for all } i = 2, 3, \ldots, n\}.$$

We will prove that the D_i are convex and $S_i \subset D_i$. Fix k, for instance $k = 1$. Define

$$H_i = \{X \in \mathbb{R}^2; h_k(X) \le h_i(X)\}.$$

Then we claim that H_i are half-planes. Let $i = 2$ for simplicity. Choose the point $P \in |O_1 O_2|$ such that $|PO_1|^2 - r_1^2 = |PO_2|^2 - r_2^2$. Take for instance $|PO_1| = \frac{|O_1 O_2|^2 - (r_1^2 - r_2^2)}{2|O_1 O_2|}$. If l is the line perpendicular to $O_1 O_2$ at P and $X \in l$, then we have $|XO_1|^2 - r_1^2 = |PO_1|^2 + |PX|^2 - r_1^2 = |PO_2|^2 + |PX|^2 - r_2^2 = |XO_2|^2 - r_2^2$. If $\tilde{H}_2$ is the half-plane that contains O_1, then $|XO_1|^2 - r_1^2 \le |XO_2|^2 - r_2^2$, for all $X \in \tilde{H}_2$, $|XO_1|^2 - r_1^2 \ge |XO_2|^2 - r_2^2$ for all $X \notin \tilde{H}_2$. Therefore, H_2 is the half-plane $\tilde{H}_2$. Consequently,

$$D_k = M \bigcap_{i=1, i \ne k}^{n} H_n$$

is a convex set as an intersection of convex sets, and $S_k \subset D_k$. It is clear that $\bigcup D_i = M$.

Problem 3.54. Consider an inscribable n-gon partitioned by means of $n-2$ nonintersecting diagonals into $n-2$ triangles. Prove that the sum of the radii of the circles inscribed in these triangles does not depend on the particular partition.

Solution 3.54. If we have a triangle inscribed in a circle of center O and radius R, of inradius r, then the distances from O to the sides satisfy the well-known formula

$$\sum_{\text{cyclic}} d^*(O, \text{side}) = R + r.$$

Here d^* is the signed distance, which is positive on one half-plane determined by the side and negative on the other half-plane. We write the corresponding formulas for each triangle of the partition and sum up all the terms. Let $d_1, \ldots, d_n$ be the distances from O to the sides of the n-gon. When looking at the signed distances from O to a diagonal, we observe that each signed distance is considered twice: once with the plus sign and once with the opposite sign. When these terms are summed, they will cancel. We then obtain

$$d_1 + \cdots + d_n = (n-2)R + r_1 + \cdots + r_n,$$

and thus the sum of inradii $r_1 + r_2 + \cdots + r_n$ does not depend on the partition.

Problem 3.55. Prove that in an ellipse having semiaxes of lengths a and b and total length L, we have $L > \pi(a + b)$.

Solution 3.55. We cut the ellipse along the axes into four congruent pieces. We can further rearrange the four pieces by adding a square of side length $b - a$ as in the figure below:

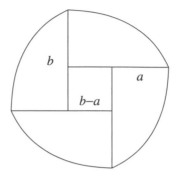

We want to use the well-known Bonnesen's isoperimetric inequality, which states that

$$L^2 \geq 4\pi A$$

for any planar convex domain of perimeter L and area A.

The area of the new domain is $A + (b - a)^2$, where A was the area of the ellipse. Moreover, one knows that the area of the ellipse is $A = \pi ab$. Applying this inequality to the domain pictured in the figure above, we obtain

$$L^2 > 4\pi^2 ab + 4\pi(b - a)^2 > \pi^2(a + b)^2,$$

whence the claim.

Problem 3.56. Let F be a convex planar domain and F' denote its image by a homothety of ratio $-\frac{1}{2}$. Is it true that one can translate F' in order for it to be contained in F? Can the constant $\frac{1}{2}$ be improved? Generalize to n dimensions.

Solution 3.56. Let $ABC \subset F$ be a triangle of maximal area inscribed in F. Let O denote the geometric centroid of ABC. We use the homothety of center O and ratio $-\frac{1}{2}$ that transforms F into F' and ABC into $A'B'C'$. Since O is the centroid of ABC, we find that A' is the midpoint of the segment $|BC|$, B' is the midpoint of $|CA|$, and C' is the midpoint of $|AB|$.

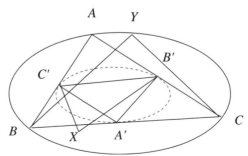

We will show that $F' \subset ABC$. Assume the contrary, namely that there exists a point X of F' that does not belong to the triangle ABC. By symmetry, it suffices to consider that X belongs to the half-plane determined by the line BC, which does not contain A. Since BC and $B'C'$ are parallel, it follows that the area of $XB'C'$ is strictly greater than the area of $A'B'C'$. Let us consider the inverse image Y of X by the homothety of center O and ratio $-\frac{1}{2}$. Then Y belongs to F and the area of YBC is strictly greater than the area of ABC, which contradicts our assumptions.

This proves that $F' \subset ABC \subset F$. The constant $\frac{1}{2}$ is sharp in the case that F is a triangle.

The same proof shows that the image F' of a convex domain F in $\mathbb{R}^n$ by the homothety of ratio $-\frac{1}{n}$ can be translated in order to be contained in F. The constant $\frac{1}{n}$ is sharp, with equality for the simplex.

Problem 3.57. A classical theorem, due to Cauchy, states that a strictly convex polyhedron in $\mathbb{R}^3$ whose faces are rigid, must be globally rigid. Here, rigidity means continuous rigidity, in the sense that any continuous deformation of the polyhedron in $\mathbb{R}^3$ that keeps the lengths of edges fixed is the restriction of a deformation of rigid Euclidean motions of three-space. Prove that a 3-dimensional cube immersed in $\mathbb{R}^n$ remains rigid for all $n > 3$.

Solution 3.57. Three sides are common to a vertex and determine a three-dimensional space $\mathbb{R}^3 \subset \mathbb{R}^n$. The 3-dimensional spaces determined by two adjacent vertices have two sides in common, and therefore they coincide. Thus, the entire cube is immersed in a 3-dimensional subspace of $\mathbb{R}^n$, and therefore it remains rigid.

The same argument shows that a rigid polyhedron with the property that exactly three sides meet at any vertex remains rigid after immersion in $\mathbb{R}^n$, for all $n > 3$.

Problem 3.58. Consider finitely many great circles on a sphere such that not all of them pass through the same point. Show that there exists a point situated on exactly two circles. Deduce that if we have a set of n points in the plane, not all of them lying on the same line, then there must exist one line passing through precisely two points of the given set.

Solution 3.58. Assume that the contrary holds. Consider the partition of the sphere into (spherical) polygons, induced by the circles, and identify it with a polyhedron. Each vertex will have degree at least 6, since at least three great circles pass through each point. If V is the number of vertices, E the number of edges, and F the number of faces of the spherical polyhedron, then Euler's formula states that

$$V - E + F = 2.$$

Moreover, by counting the edges in each face and summing over the faces, we obtain

$$2E = 3F_3 + 4F_4 + \cdots \geq 3F,$$

where F_k is the number of faces with k sides. Finally, counting the edges adjacent to each vertex and summing over the vertices, we get

$$2E = 6V_6 + 7V_7 + \cdots \geq 6V,$$

where V_k is the number of vertices of degree k, because $V_k = 0$ when $k \leq 5$. Thus $6(E + 2) = 6(V + F) \leq 2E + 4E = 6E$, which is absurd. Therefore, there exists a vertex of valence smaller than 6 in which exactly two great circles intersect.

The second claim can be reformulated using projective duality: if we have n lines in the projective plane, not all passing through the same point, then there exists a point lying on exactly two of the lines. This is equivalent to the problem considered above.

Comments 72 *The problem was posed by Sylvester in 1839, and it was not settled until 1930, when T. Gallai gave a solution. The solution above is due to N. Steenrod. Moreover, it can be proved that the number of points where exactly two circles intersect is at least $3n/7$; see for instance:*

- H.S.M. Coxeter: *The classification of zonohedra by means of projective diagrams,* J. Math. Pures Appl. (9) 41 (1962), 137–156.
- G.D. Chakerian, *Sylvester problem on collinear points and a relative,* Amer. Math. Monthly 77 (1970), 164–167.

Problem 3.59. Given a finite set of points in the plane labeled with $+1$ or -1, and not all of them collinear, show that there exists a line determined by two points in the set such that all points of the set lying on that line are of the same sign.

Solution 3.59. We use projective duality as above and reformulate the question as follows: given n great circles on a sphere, to each one being assigned either $+1$ or -1, and not all passing through the same point, show that there exists a point lying on at least two circles such that all circles containing that point have the same sign.

The circles determine a spherical polyhedron whose edges are labeled by $+1$ and -1. Assume that the result is not true. Then the edges around each vertex must have both the labels $+1$ and -1. By symmetry, the number of sign changes in the labels of the edges that one meets when traveling-cyclically around each vertex is at least 4.

Denote by N the sum of all such sign changes over all vertices. Then $N \geq 4V$.

On the other hand, by counting the sign changes as we travel around the faces and summing over all faces, we should again obtain N. Further, the number of sign changes in a k-sided face is even and at most k, so that

$$N \leq 2F_3 + 4F_4 + 4F_5 + 6F_6 + 6F_7 + \cdots .$$

Moreover, using Euler's formula and

$$2E = 3F_3 + 4F_4 + \cdots$$

we obtain

$$4V - 8 = 4E - 4F = 2(3F_3 + 4F_4 + 5F_5 + \cdots) - 4(F_3 + F_4 + \cdots)$$
$$= 2F_3 + 4F_4 + 6F_5 + \cdots \geq N \geq 4V,$$

which is a contradiction. Thus our claim follows.

Comments 73 *The result concerning the sign changes is valid more generally when some edges are allowed to be unlabeled, and as such it is known as Cauchy's lemma. This is an essential ingredient in proving the famous rigidity theorem of Cauchy, which states that a convex polyhedron in $\mathbb{R}^3$ whose faces are rigid is globally rigid.*

Problem 3.60. If Q is a given rectangle and $\varepsilon > 0$, then Q can be covered by the union of a finite collection S of rectangles with sides parallel to those of Q in such a way that the union of every nonoverlapping subcollection of S has area less than ε.

Solution 3.60. Observe first that every rectangle Q has a subset H that is the union of a finite collection Σ of rectangles with parallel sides such that:

1. any two members of Σ overlap;
2. area$(H) > \delta$ area(Q), where $\delta = \exp(-1/\varepsilon)$;
3. area$(A) < \frac{\varepsilon}{2}$ area(H), for any $A \in \Sigma$.

If $Q = [0, a] \times [0, b]$, we consider rectangles A_i that are parallel and have one vertex at $(0, 0)$ and the other vertex on the curve of equation $xy = \delta$ area(Q). Thus each rectangle has area δ. The curved triangle determined by the inequalities $xy \le \delta$ area(Q), x, $y \ge 0$, will be denoted by T. Moreover, we consider a collection of such rectangles such that their union H satisfies area$(H) > \frac{1+\varepsilon}{\varepsilon}$ area$(T) = \delta$ area(Q). Further, area$(A) = \delta$ area$(Q) < \varepsilon$ area(T).

Set $Q_0 = Q$, $H_0 = H$, $\Sigma_0 = \Sigma$ and assume that area$(Q) = 1$, for simplicity. Then $Q_0 - H_0$ is a finite union of parallel rectangles $Q_{i,0}$ and

$$\text{area}(Q_0 - H_0) < (1 - \delta).$$

Choose in each rectangle Q_i a set of type $H_{i,0}$ as above. The union of all such $H_{i,0}$ and H_0 will be denoted by H_1, and the collection of rectangles will be denoted by Σ_1. Then $Q - H_1$ is also a finite union of parallel rectangles $Q_{i,1}$ and

$$\text{area}(Q - H_1) < (1 - \delta)^2 \text{ area}(Q).$$

We continue this process by defining the sets H_j and the collections Σ_j of rectangles such that $Q - H_j$ is a finite union of rectangles $Q_{i,j}$ whose complement has total area

$$\text{area}(Q - H_j) < (1 - \delta)^j \text{ area}(Q).$$

We stop this process at the stage n, where n is large enough that $(1 - \delta)^n$ area$(Q) < \varepsilon/2$. Consider then the union Σ of the collections Σ_n obtained so far with the set of rectangles in which $Q - H_n$ decomposes.

If we consider a subfamily of disjoint rectangles from Σ, then there is at most one element A_i from any collection $\Sigma_{i,j}$ covering $H_{i,j}$, because all elements in $\Sigma_{i,j}$ overlap each other. Thus the total area covered by these A_i is at most

$$\sum_{i,j} \frac{\varepsilon}{2} \text{ area}(H_{i,j}) < \frac{\varepsilon}{2}.$$

Since the rectangles from $Q - H_n$ contribute at most $\frac{\varepsilon}{2}$, we are done.

Problem 3.61. Prove that the 3-dimensional ball cannot be partitioned into three sets of strictly smaller diameter.

Solution 3.61. Assume that we can divide the unit ball into three pieces of smaller diameter and let X_1, X_2, X_3 be the partition that it induces on the boundary sphere. Thus the diameter of each piece X_i is smaller than $1 - 10h$, for some $h > 0$. We choose a very thin triangulation of the sphere with triangles that have edges smaller than h. Denote by N_j the union of those spherical triangles of the triangulation above that intersect the set X_j nontrivially. It follows that N_1, N_2, N_3 are closed subsets of the sphere that are domains, i.e., they are closures of open subsets. In particular, the boundary of each domain N_j is the union of several piecewise linear circles. Moreover, the union $N_1 \cup N_2 \cup N_3$ covers the sphere.

Since each triangle has a small size, the diameter of N_j is at most $1 - 8h < 1$. In particular, opposite points cannot belong simultaneously to the same set N_j. Let us denote by x^* the point of the sphere opposite the point x. We know that the set N_1 and its symmetric N_1^* (which is the set of those points x^* for which $x \in N_1$) should be disjoint; otherwise, there would exist in N_1 a pair of points at distance 1. Consider the circles determined by the boundary circles of both N_1 and N_1^*. The circles associated to N_j are disjoint and distinct from the circles associated to N_1^*, because $N_1 \cap N_1^* = \emptyset$. Thus, there is an even number of such circles. Observe that k disjoint circles on the sphere divide the surface of the sphere into $k + 1$ connected complementary regions. Thus, the collection of boundary circles associated to N_1 and N_1^* will split the surface of the sphere into an odd number of connected regions, which we denote by $R_1, R_2, \ldots, R_{2m+1}$.

Consider the symmetric R_j^* of the region R_j. There are two possibilities: either R_j^* is another connected region R_k that is disjoint from R_j, or else $R_j^* = R_j$, meaning that R_j is symmetric. The second case happens, for instance, when the region R_j is bounded by two circles, one coming from N_1 and the other from N_1^*. If there were no symmetric regions R_j, then the total number of regions would be even, which would contradict our previous claim. Thus there must be at least one connected symmetric region R_j.

We consider a point $x \in R_j$ and its symmetric x^*, which must also belong to R_j, since the region is symmetric. If $x \in N_1$, then all points of the sphere that are connected to x by a path not hitting a boundary circle of N_1 must also belong to N_1. Thus, the connected component R_j will be contained in N_1, and so N_1 will contain pairs of symmetric points, contradiction. This implies that $x \notin N_1$ and hence $x \in N_2 \cup N_3$. Let us assume that $x \in N_2$. Then its symmetric x^* is not in N_2, since N_2 has diameter smaller than 1 and thus $x^* \in N_3$. Further, R_j is connected and thus there exists a path γ within R_j that connects x to x^*. One knows that N_2 and N_3 are domains covering the path γ. Moreover, there exists a point $z \in \gamma$ that belongs to both N_2 and N_3. In fact, consider the point $z \in \gamma$ closest to x^* and that belongs to N_2. There exists such a point since N_2 is closed. If $z = x^*$, then both x and x^* belong to N_2, contradiction. If $z \neq x^*$, then the points on the right of z should belong to N_3 and thus z belongs to the closure of N_3 and thus to N_3.

Now $z \in R_j$ and thus its symmetric z^* also belongs to R_j. This means that $z, z^* \notin N_1$. But z^* can belong neither to N_2, since N_2 does not contain opposite points, nor to N_3, for the same reason. This contradiction proves the claim.

Comments 74 *More generally, it is known that the unit n-ball (or equivalently, the $(n-1)$-sphere in $\mathbb{R}^n$) cannot be partitioned into n pieces of strictly smaller diameter. This is related to Borsuk's problem discussed in comments to Problem 3.38. The result was proved by Lusternik and Shnirelman in 1930 and, independently, by Borsuk in 1932, but the higher-dimensional case $n \geq 4$ requires deeper methods from topology, and this opened a new chapter in algebraic topology. The interested reader might consult the survey of Steinlein, which presents both historical remarks and a large number of references.*

- K. Borsuk: *Über die Zerlegung einer n-dimensionalen Vollkugel in n-Mengen,* Verh. International Math. Kongress Zürich 1932, 192.
- L.A. Lusternik and L.G. Shnirelman: *Topological Methods for Variational Problems,* Moscow, 1930.
- H. Steinlein: *Borsuk's antipodal theorem and its generalizations and applications: a survey,* Topological Methods in Nonlinear Analysis, 166–235, Sém. Math. Sup., 95, Presses Univ. Montreal, Montreal, 1985.
- H. Steinlein: *Spheres and symmetry: Borsuk's antipodal theorem.* Topol. Methods Nonlinear Anal. 1 (1993), 1, 15–33.

7.3 Geometric Inequalities

Problem 3.62. If a, b, c, r, R are the usual notations in the triangle, show that

$$\frac{1}{2rR} \leq \frac{1}{3}\left(\sum \frac{1}{a}\right)^2 \leq \sum \frac{1}{a^2} \leq \frac{1}{4r^2}.$$

Solution 3.62. We derive from $\sum(\frac{1}{a} - \frac{1}{b})^2 \geq 0$ that

$$\sum \frac{1}{bc} \leq \frac{1}{3}\left(\sum \frac{1}{a}\right)^2 \leq \sum \frac{1}{a^2}.$$

Moreover, the usual Cauchy–Schwarz inequality shows that $(p-a)(p-b) \leq \frac{c^2}{4}$, and its analogues. Then we have

$$\frac{1}{2rR} = \frac{2p}{abc} = \frac{a+b+c}{abc} = \frac{1}{ab} + \frac{1}{bc} + \frac{1}{ca} \leq \frac{1}{3}\left(\sum \frac{1}{a}\right)^2 \leq \sum \frac{1}{a^2}$$

$$\leq \frac{1}{4}\left(\sum \frac{1}{(p-b)(p-c)}\right) = \frac{p}{4(p-a)(p-b)(p-c)} = \frac{1}{4r^2}.$$

Problem 3.63. If a, b, c are the sides of a triangle, then prove that $\frac{(b+c)^2}{4bc} \leq \frac{m_a}{w_a}$ and $\frac{b^2+c^2}{2bc} \leq \frac{m_a}{k_a}$, where m_a, w_a, k_a denotes the respective lengths of the median, bisector, and altitude issued from A.

Solution 3.63. We have the following formulas computing m_a, w_a, and k_a:

$$m_a = \frac{1}{2}\sqrt{2(b^2 + c^2) - a^2}, \quad w_a = \frac{\sqrt{bc}}{b + c}\sqrt{(b + c)^2 - a^2},$$

$$k_a = \frac{bc}{b^2 + c^2}\sqrt{2(b^2 + c^2) - a^2}.$$

From the triangle inequality, we have $a^2 > (b - c)^2$, which implies furthermore that

$$(b + c)^2 - a^2 < 4bc \frac{(b - c)^2}{4bc} \leq \frac{(b - c)^2}{(b + c)^2 - a^2},$$

with equality only if $b = c$. Adding 1 in both members yields

$$\frac{(b + c)^2}{4bc} \leq \frac{2b^2 + 2c^2 - a^2}{(b + c)^2 - a^2}; \text{ hence } \frac{b + c}{2\sqrt{bc}} \leq \sqrt{\frac{2b^2 + 2c^2 - a^2}{(b + c)^2 - a^2}}.$$

This leads to

$$\frac{(b + c)^2}{4bc} \leq \frac{b + c}{2\sqrt{bc}}\sqrt{\frac{2(b^2 + c^2) - a^2}{(b + c)^2} - a^2} = \frac{m_a}{w_a}.$$

For the second inequality, it suffices to observe that

$$\frac{m_a}{h_a} \geq \frac{m_a}{k_a} = \frac{b^2 + c^2}{2bc}.$$

Notice that equality in the first case implies $b = c$, while for the second case either $b = c$ or $a^2 = b^2 + c^2$.

Problem 3.64. If $S(x, y, z)$ is the area of a triangle with sides x, y, z, prove that

$$\sqrt{S(a, b, c)} + \sqrt{S(a', b', c')} \leq \sqrt{S(a + a', b + b', c + c')}.$$

Solution 3.64. We make the following change of variables: $s = (a + b + c)/2$, $t = s - a$, $u = s - b$, and $v = s - c$. Using Heron's formula, the inequality becomes

$$\sqrt[4]{stuv} + \sqrt[4]{s't'u'v'} \leq \sqrt[4]{(s + s')(t + t')(u + u')(v + v')}.$$

We prove first that for positive x, x', y, y', we have

$$\sqrt{xy} + \sqrt{x'y'} \leq \sqrt{(x + x')(y + y')}.$$

In fact, this follows from $xy + x'y' + 2\sqrt{x'y \cdot y'x} \leq xy + x'y' + xy' + yx'$, and equality holds when the geometric means of $x'y$ and $y'x$ are equal.

By applying the last inequality to $x = \sqrt{st}$ and $y = \sqrt{uv}$ twice, we obtain the claim. Moreover, the equality is attained only when $t/t' = u/u' = v/v' = s/s'$, which amounts to $a/a' = b/b' = c/c'$.

Problem 3.65. It is known that in any triangle we have the inequalities

$$3\sqrt{3}r \le p \le 2R + (3\sqrt{3} - 4)r,$$

where p denotes the semiperimeter. Prove that in an obtuse triangle, we have

$$(3 + 2\sqrt{2})r < p < 2R + r.$$

Solution 3.65. We have

$$\frac{a+b}{c} = \frac{\cos \frac{A-B}{2}}{\sin \frac{C}{2}}.$$

Since $C > \frac{\pi}{2}$, we have $\sin \frac{C}{2} > \frac{\sqrt{2}}{2}$ and thus $c\sqrt{2} > a + b$. Therefore

$$2p\sqrt{2} > (a+b)(\sqrt{2}+1) \Rightarrow 8p^2$$
$$> (3 + 2\sqrt{2})(a+b)^2 > 4(3 + 2\sqrt{2})ab$$
$$> 8(3 + 2\sqrt{2})S,$$

where S is the area. This implies that

$$p > \frac{(3 + 2\sqrt{2})S}{p} = (3 + 2\sqrt{2})r.$$

Finally, recall that $r = (p - c) \tan \frac{C}{2} > (p - c)$ and so $p < c + r = 2R \sin C + r < 2R + r$.

Problem 3.66. Prove the Euler inequality

$$R \ge 2r.$$

Solution 3.66.

$$|OI|^2 = R^2 - 2Rr.$$

Problem 3.67. Prove that in a triangle we have the inequalities

$$36r^2 \le a^2 + b^2 + c^2 \le 9R^2.$$

Solution 3.67. We have

$$\frac{p}{3} = \frac{1}{3}(p - a + p - b + p - c) \ge \sqrt[3]{(p-a)(p-b)(p-c)} = \sqrt[3]{\frac{S^2}{p}} = \sqrt[3]{r^2 p},$$

and hence $p^2 \ge 27r^2$. Thus

$$a^2 + b^2 + c^2 \ge \frac{4}{3}p^2 \ge 36r^2.$$

The second inequality is a consequence of the following identity:

$$|OH|^2 = 9R^2 - (a^2 + b^2 + c^2),$$

where O is the circumcenter and H the orthocenter.

Problem 3.68.   1. Let ABC and $A'B'C'$ be two triangles. Prove that

$$\frac{a^2}{a'} + \frac{b^2}{b'} + \frac{c^2}{c'} \le R^2 \frac{(a'+b'+c')^2}{a'b'c'}.$$

2. Derive that

$$a^2 + b^2 + c^2 \le 9R^2,$$

$$\cos A \cos B \cos C \le \frac{1}{8}.$$

Solution 3.68. 1. The inequality from the statement is equivalent to

$$a'^2 + b'^2 + c'^2 - 2(b'c' \cos A'' + a'c' \cos B'' + b'a' \cos C'') \ge 0,$$

where $A'' = \pi - 2A$, $B'' = \pi - 2B$, $C'' = \pi - 2C$. Moreover, this can be written as

$$(a' - b' \cos C'' - c' \cos B'')^2 + (b' \sin C'' - c' \sin B'')^2 \ge 0.$$

A necessary condition for equality is the vanishing of the second square, i.e.,

$$\frac{a'}{\sin A''} = \frac{b'}{\sin B''} = \frac{c'}{\sin C''}.$$

But $A'' \in (-\pi, \pi)$; therefore A'', B, C'' are the angles of a triangle that is similar to $A'B'C'$. Therefore, we obtain equality if and only if ABC is an acute triangle and $A'B'C'$ is similar to the orthic triangle of the triangle ABC.

2. If $A'B'C'$ is equilateral, then $a^2 + b^2 + c^2 \le 9R^2$, which is equivalent to $\sin^2 A + \sin^2 B + \sin^2 C \le \frac{9}{4}$. Observe that $\sum \sin^2 A = 2 + 2 \cos A \cos B \cos C$, and therefore we obtain $\cos A \cos B \cos C \le \frac{1}{8}$.

Problem 3.69. Prove that the following inequalities hold in a triangle:

$$4 \sum_{\text{cyclic}} h_A h_B \le 12S\sqrt{3} \le 54Rr \le 3 \sum_{\text{cyclic}} ab \le 4 \sum_{\text{cyclic}} r_A r_B.$$

Solution 3.69. 1. We have

$$4 \sum_{\text{cyclic}} r_A r_B = \left(\sum_{\text{cyclic}} a \right)^2 \ge 3 \sum_{\text{cyclic}} ab.$$

2. In order to prove that

$$18Rr \le \sum_{\text{cyclic}} ab,$$

we start from the identity

$$\frac{1}{h_a} + \frac{1}{h_b} + \frac{1}{h_c} = \frac{1}{r}.$$

Thus

$$2\frac{abc}{4R}\left(\frac{1}{a}+\frac{1}{b}+\frac{1}{c}\right)=2S\left(\frac{1}{a}+\frac{1}{b}+\frac{1}{c}\right)=h_a+h_b+h_c\geq\frac{9}{\sum\frac{1}{h_a}}\geq 9r,$$

leading to the inequality we wanted.

3. Recall that Jensen's inequality states that if f is a concave (i.e., $f''(x)\leq 0$) smooth real function on some interval J, then

$$\lambda_1 f(x_1)+\lambda_2 f(x_2)+\cdots+\lambda_n f(x_n)\leq f(\lambda_1 x_1+\lambda_2 x_2+\cdots+\lambda_n+x_n)$$

for all $x_j\in J$ and $\lambda_i\geq 0$ such that $\lambda_1+\lambda_2+\cdots+\lambda_n=1$.

Notice now that the function cos is concave on $\left[0,\frac{\pi}{2}\right]$, and by Jensen's inequality we have

$$\cos\frac{A}{2}+\cos\frac{B}{2}+\cos\frac{C}{2}\leq\frac{3\sqrt{3}}{2}.$$

Therefore, by the means inequality,

$$\cos\frac{A}{2}\cos\frac{B}{2}\cos\frac{C}{2}\leq\left(\frac{\cos\frac{A}{2}+\cos\frac{B}{2}+\cos\frac{C}{2}}{3}\right)^3\leq\frac{3\sqrt{3}}{8}.$$

We have then

$$S=\frac{abc}{4R}=4rR\cos\frac{A}{2}\cos\frac{B}{2}\cos\frac{C}{2}\leq\frac{3\sqrt{3}}{2}rR.$$

4. We have

$$\sum_{\text{cyclic}}h_A h_B=h_A h_B h_C\left(\frac{1}{h_a}+\frac{1}{h_b}+\frac{1}{h_c}\right)=\frac{2S^2}{R}\frac{1}{r}\leq 3\sqrt{3}S.$$

Problem 3.70. Prove that in an any triangle ABC, we have

$$\frac{\sqrt{1+8\cos^2 B}}{\sin A}+\frac{\sqrt{1+8\cos^2 C}}{\sin B}+\frac{\sqrt{1+8\cos^2 A}}{\sin C}\geq 6.$$

Solution 3.70. By applying the means inequality, the left-hand side is greater than

$$\sqrt[3]{3\left(\prod_{\text{cyclic}}(\sqrt{1+8\cos^2 A})\right)\prod_{\text{cyclic}}\sqrt[3]{\frac{1}{\sin A}}},$$

and it is sufficient to prove that

$$F(A,B,C)=\prod_{\text{cyclic}}\frac{1+8\cos^2 A}{\sin^2 A}\geq 64.$$

Let us find the extremal points of F, that satisfy

$$\frac{\partial F}{\partial A} = \frac{\partial F}{\partial B} = \frac{\partial F}{\partial C}.$$

We have

$$\frac{1}{F} \cdot \frac{\partial F}{\partial A} = -\frac{18 \cot A}{1 + 8 \cos^2 A},$$

and thus we have

$$\sin B \cos C (1 + 8 \cos^2 B) = \sin C \cos B (1 + 8 \cos^2 C).$$

This implies that $\sin(B - C) + 4 \cos B \cos C (\sin 2B - \sin 2C) = 0$ and hence

$$\sin(B - C)(1 - 8 \cos A \cos B \cos C) = 0.$$

We claim that this happens only if $A = B = C$. In fact, if the triangle is obtuse, then

$$\cos A \cos B \cos C \leq 0 < \frac{1}{8}$$

and hence $B = C = A$.
   Otherwise, $\cos$ is positive and

$$\cos A + \cos B + \cos C = 1 + 4 \sin \frac{A}{2} \sin \frac{B}{2} \sin \frac{C}{2} = 1 + 4 \prod_{\text{cyclic}} \sqrt{(p - b)(p - c)}bc$$

$$= 1 + \frac{r}{R} \leq \frac{3}{2},$$

and so by the means inequality,

$$\cos A \cos B \cos C \leq \left(\frac{\cos A + \cos B + \cos C}{3} \right)^3 \leq \frac{1}{8}.$$

Equality holds above only if the triangle is equilateral.
   This shows that the only extremal points are $A = B = C = \frac{\pi}{3}$, and one verifies that it corresponds to a minimum, so that $F(A, B, C) \geq 64$, as claimed.

Problem 3.71. Let P be a point in the interior of the triangle ABC. We denote by R_a, R_b, R_c the distances from P to A, B, C and by r_a, r_b, r_c the distances to the sides BC, CA, AB. Prove that

$$\sum_{\text{cyclic}} R_a^2 \sin^2 A \leq 3 \sum_{\text{cyclic}} r_a^2,$$

with equality if and only if P is the Lemoine point (i.e., the symmedian point).

Solution 3.71. Let G be the centroid of the poder triangle DEF of P, obtained by projecting P to the sides of ABC. We have then

$$3 \sum_{\text{cyclic}} r_a^2 = 3(|DG|^2 + |EG|^2 + |FG|^2 + 3|PG|^2) \geq 3(|DG|^2 + |EG|^2 + |FG|^2)$$

$$= |DE|^2 + |EF|^2 + |FD|^2 = \sum_{\text{cyclic}} R_a^2 \sin^2 A.$$

We obtain equality when P is the centroid of its poder triangle. Therefore P is the symmedian point.

Comments 75 *We can generalize this type of inequality. We have, for instance,*

$$x R_1^2 + y R_1^2 + z R_1^2 \geq \frac{a^2 yz + b^2 xz + c^2 xy}{x + y + z}$$

for every $x, y, z \in \mathbb{R}$ such that $x + y + z > 0$. The equality holds above if and only if

$$\frac{x}{F_1} = \frac{y}{F_2} = \frac{z}{F_3},$$

where F_1, F_2, F_3 denote the areas of BPC, APC, and APB, respectively.

Comments 76 *With every inequality $\Phi(R_a, R_b, R_c, a, b, c) \geq 0$, valid for any triangle of sides a, b, c, we can associate a dual inequality in a new set of variables, $\Phi(r_a, r_b, r_c, R_a \sin A, R_b \sin B, R_c \sin C) \geq 0$, obtained by replacing the original triangle by the poder triangle associated to P. Then the distances from P to the vertices of the poder triangle are r_a, r_b, r_c and the sides of the poder triangle are $R_a \sin A, R_b \sin B, R_c \sin C$.*

For instance, the dual of the inequality from the previous remark is

$$x r_a^2 + y r_b^2 + z r_c^2 \geq \frac{yz R_a^2 \sin^2 A + xz R_b^2 \sin^2 B + xy R_c^2 \sin^2 C}{x + y + z}.$$

If $x = y = z$, then this is the inequality of our problem. Applying the inequality from the previous remark to the right-hand side, we also obtain

$$\frac{\sum yz R_1^2 \sin^2 A}{x + y + z} \geq \frac{4S^2}{\frac{a^2}{x} + \frac{b^2}{y} + \frac{c^2}{z}}.$$

Problem 3.72. Prove the inequalities

$$16Rr - 5r^2 \leq p^2 \leq 4R^2 + 4Rr + 3r^2.$$

Solution 3.72. Let I be the incenter of the triangle and H be the orthocenter. Direct calculations show that

$$|IH|^2 = 2r^2 - 4R^2 \cos A \cos B \cos C,$$

and replacing the term

$$4R^2 \cos A \cos B \cos C = p^2 - (2R + r)^2,$$

we find that

$$|IH|^2 = 4R^2 + 4Rr + 3r^2 - p^2.$$

The next step is to use Euler's theorem, which states that O, G, and H are collinear and

$$\frac{|OG|}{|GH|} = \frac{1}{2},$$

where G denotes the centroid of the triangle. We compute the length of $|GI|$ using the triangle HIO as follows:

$$|GI|^2 = \frac{2}{3}|IO|^2 + \frac{1}{3}|IH|^2 + \frac{2}{9}|OH|^2.$$

We substitute

$$|OI|^2 = R^2 - 2Rr, \quad |OH|^2 = 9R^2 - (a^2 + b^2 + c^2)$$

and derive that

$$|GI|^2 = r^2 - \frac{1}{3}p^2 + \frac{2}{9}(a^2 + b^2 + c^2).$$

Moreover, $4p^2 = 2(a^2 + b^2 + c^2) + 16Rr + 4r^2$, and so

$$p^2 - 16Rr + 5r^2 = 8|GI|^2.$$

Thus the inequality follows.

Comments 77 *Using $2p^2 = 2r(4R + r) + a^2 + b^2 + c^2$ and $S = rp$, we derive also the inequalities*

$$12r(2R - r) \le a^2 + b^2 + c^2 \le 4r^2 + 8R^2,$$
$$r^3(16R - 5r) \le S^2 \le r^2(3r^2 + 4rR + 4R^2).$$

- O. Bottema, R.Ž. Djordjević, R.R. Janić, D.S. Mitrinović, and P.M. Vasić: *Geometric Inequalities,* Wolters–Noordhoff Publishing, Gröningen 1969.

Problem 3.73. Prove the following inequalities, due to Roché:

$$2R^2 + 10Rr - r^2 - 2(R - 2r)\sqrt{R^2 - 2Rr}$$
$$\le p^2 \le 2R^2 + 10Rr - r^2 + 2(R - 2r)\sqrt{R^2 - 2Rr}.$$

Solution 3.73. Suppose that $a \ge b \ge c$. By direct calculation, we compute the area of this triangle as follows:

$$\text{area}(HIO) = 2R^2 \sin\left(\frac{B - C}{2}\right) \sin\left(\frac{A - C}{2}\right) \sin\left(\frac{A - B}{2}\right) = \frac{(b - c)(a - c)(a - b)}{8r}.$$

Expressing R and r in terms of a, b, c, we can conclude with the identity

$$(\text{area}(HIO))^2 = -p^4 + 2(2R^2 + 10Rr - r^2)p^2 - r(4R + r)^3.$$

Since this binomial in p^2 is positive, the value of p^2 lies between the two real roots of the polynomial, which yields the claim.

Comments 78 *W.J. Blundon gave another proof, using the following identity:*

$$-p^4 + 2(2R^2 + 10Rr - r^2)p^2 - r(4R + r)^3 = \frac{1}{4r^2}(a - b)^2(b - c)^2(c - a)^2.$$

- D.S. Mitrinović, J.E. Pečarić, V. Volenec: *Recent Advances in Geometric Inequalities,* Math. Appl. (East European Series), 28, Kluwer Academic Publ., Dordrecht, 1989.

8

Analysis Solutions

Problem 4.1. Prove that $z \in \mathbb{C}$ satisfies $|z| - \Re z \leq \frac{1}{2}$ if and only if $z = ac$, where $|\bar{c} - a| \leq 1$. We denote by $\Re z$ the real part of the complex number z.

Solution 4.1. We have the identity

$$|ac| - \Re \, ac = \frac{1}{2}|\bar{c} - a|^2 - \frac{1}{2}(|c| - |a|)^2,$$

and the "if" part follows. For the converse, we consider a, c such that $|a| = |c| = |z|^{1/2}$.

Problem 4.2. Let $a, b, c \in \mathbb{R}$ be such that $a + 2b + 3c \geq 14$. Prove that $a^2 + b^2 + c^2 \geq 14$.

Solution 4.2. From $(a - 1)^2 + (b - 2)^2 + (c - 3)^2 \geq 0$, we obtain $a^2 + b^2 + c^2 \geq 2(a + 2b + 3c) - 14 \geq 14$.

Comments 79 *More generally, if $w_i \in \mathbb{R}_+$ and $a_i \in \mathbb{R}$ are such that $\sum_{i=1}^{n} a_i w_i \geq \sum_{i=1}^{n} w_i^2$, then $\sum_{i=1}^{n} a_i^2 \geq \sum_{i=1}^{n} w_i^2$.*

Problem 4.3. Let $f_n(x)$ denote the Fibonacci polynomial, which is defined by

$$f_1 = 1, \quad f_2 = x, \quad f_n = xf_{n-1} + f_{n-2}.$$

Prove that the inequality

$$f_n^2 \leq \left(x^2 + 1\right)^2 \left(x^2 + 2\right)^{n-3}$$

holds for every real x and $n \geq 3$.

Solution 4.3. Since $f_3(x) = x^2 + 1$, the inequality is trivially satisfied for $n = 3$. We proceed by induction on n:

$$f_{n+1}^2(x) = [xf_n(x) + f_{n-1}(x)]^2$$
$$\leq [x(x^2+1)(x^2+2)^{(n-3)/2} + (x^2+1)(x^2+2)^{(n-4)/2}]^2$$
$$\leq (x^2+1)^2(x^2+2)^{n-2}[x(x^2+2)^{-1/2} + (x^2+2)^{-1}]^2$$
$$< (x^2+1)^2(x^2+2)^{k-2}.$$

We have used above that $x(x^2+2)^{-1/2} + (x^2+2)^{-1} < 1$, which is a consequence of $x^4 + 2x^2 < (x^2+1)^2$.

Second proof. One proves by induction that

$$f_n(x) = \det \left. \begin{pmatrix} x & -1 & 0 & 0 & \cdots & 0 & 0 \\ 1 & x & -1 & 0 & \cdots & 0 & 0 \\ 0 & 1 & x & -1 & \cdots & 0 & 0 \\ \vdots & \vdots & \vdots & \vdots & & \vdots & \vdots \\ 0 & 0 & 0 & 0 & \cdots & 1 & x \end{pmatrix} \right\} n-1.$$

Let us recall now the Hadamard inequality, which gives an upper bound for the determinant of an arbitrary $k \times k$ matrix, as follows:

$$(\det(a_{ij}))^2 \leq \prod_{j=1}^{k} \left(\sum_{i=1}^{k} a_{ij}^2 \right).$$

This yields the claimed inequality.

Problem 4.4. Prove the inequality

$$\min\left((b-c)^2, (c-a)^2, (a-b)^2\right) \leq \frac{1}{2}(a^2+b^2+c^2).$$

Generalize to $\min_{1 \leq k < i \leq n}(a_k - a_i)^2$.

Solution 4.4. We will prove that for any real numbers a_i, we have

$$\min_{i<j}(a_i - a_j)^2 \leq \mu^2(a_1^2 + \cdots + a_n^2), \text{ where } \mu^2 = \frac{12}{n(n^2-1)}.$$

There is no loss of generality if we suppose that $a_1 \leq \cdots \leq a_n$ and $\sum_{i=1}^{n} a_i^2 = 1$. Then

$$\sum_{1 \leq i < j \leq n} (a_i - a_j)^2 = n \sum_{i=1}^{n} a_i^2 - \left(\sum_{i=1}^{n} a_i \right)^2.$$

If $a_{i+1} - a_i > \mu > 0$ for all $i \leq n-1$, then $(a_j - a_i)^2 > (i-j)^2\mu^2$ and hence

$$\sum_{1 \leq i < j \leq n} (a_i - a_j)^2 > \mu^2 \sum_{1 \leq i < j \leq n} (i-j)^2 = \mu^2 \frac{n^2(n^2-1)}{12} = n.$$

This implies that

$$n < n \sum_{i=1}^{n} a_i^2 - \left(\sum_{i=1}^{n} a_i\right)^2 \leq n \sum_{i=1}^{n} a_i^2 = n,$$

which is absurd. Therefore, there exist $i \neq j$ such that $(a_i - a_j)^2 \leq \mu^2$.

Problem 4.5. Let $a_1, \ldots, a_n$ be the lengths of the sides of a polygon and P its perimeter. Then

$$\frac{a_1}{P - a_1} + \frac{a_2}{P - a_2} + \cdots + \frac{a_n}{P - a_n} \geq \frac{n}{n - 1}.$$

Solution 4.5. Set $P - a_i = y_i$. Then $\sum_{i=1}^{n} y_i = nP - \sum_{i=1}^{n} a_i = (n-1)P$. Now we express $a_i = P - y_i = \left(\frac{1}{n-1} \sum_{j=1}^{n} y_j\right) - y_i = \frac{1}{n-1}\left(\sum_{j\neq i} y_j - (n-2)y_i\right)$. Thus we can write

$$\sum_{i=1}^{n} \frac{a_i}{S - a_i} = \frac{1}{n-1} \sum_{j=1}^{n} \sum_{j\neq i} \left(\frac{y_j}{y_i} - \frac{n-2}{n-1}\right).$$

Further,

$$\sum_{j=1}^{n} \sum_{j\neq i} \left(\frac{y_j}{y_i} - \frac{n-2}{n-1}\right) = \sum_{j=1}^{n} \sum_{i<j} \left(\frac{y_j}{y_i} + \frac{y_i}{y_j} - 2\frac{n-2}{n-1}\right) \geq \frac{2n}{n-1},$$

where we have regrouped the terms with indices i, j together with those with indices j, i and used the fact that $\frac{y_j}{y_i} + \frac{y_i}{y_j} \geq 2$.

Problem 4.6. Assume that a, b, and c are positive numbers that cannot be the sides of a triangle. Then the following inequality holds

$$(abc)^2 (a + b + c)^2 (a + b - c)(a - b + c)(b - a + c)$$
$$\geq \left(a^2 + b^2 + c^2\right)^3 \left(a^2 + b^2 - c^2\right)\left(a^2 - b^2 + c^2\right)\left(b^2 - a^2 + c^2\right).$$

Solution 4.6. Let us note the following identity:

$$(a + b + c)(a - b + c)(a + c - b)(-a + b + c) = -\sum a^4 + 2\sum a^2 b^2,$$

where all sums are cyclic sums over the three letters involved. One multiplies both members of the identity by $\sum a^2$ and obtains

$$\left(a^2 + b^2 - c^2\right)(a + b + c)(a - b + c)(-a + b + c)(a + b - c)$$
$$= -\sum a^6 + \sum a^4 b^2 + 6a^2 b^2 c^2$$
$$= \left(-a^2 + b^2 + c^2\right)\left(a^2 - b^2 + c^2\right)\left(a^2 + b^2 - c^2\right).$$

This help us to write the inequality above in the following form:

$$2p(2p - a)(2p - b)(2p - c)\left((abc)^2(2p)^2 - (a^2 + b^2 + c^2)^4\right)$$
$$+ 8(abc)^2(a^2 + b^2 + c^2)^3 \geq 0,$$

where $2p = a + b + c$. Let us note that $(abc)^2(a + b + c)^2 - (a^2 + b^2 + c^2)^4 \leq 0$. In fact,

$$abc(a + b + c) = a^2bc + b^2ca + c^2ab \leq \frac{1}{2}\sum a^2(b^2 + c^2)$$
$$= \frac{1}{2}\left(\left(\sum a^2\right)^2 - \sum a^4\right) \leq (a^2 + b^2 + c^2)^2.$$

Finally, if a, b, c do not determine a triangle, then

$$2p(2p - a)(2p - b)(2p - c) \leq 0,$$

and our inequality follows.

Problem 4.7. Prove that $\sin^2 x < \sin x^2$, for $0 < x \leq \sqrt{\frac{\pi}{2}}$.

Solution 4.7. For $1 \leq x^2 \leq \frac{\pi}{2}$, the inequality is obvious. For $0 < x \leq 1$, we have $0 < t^2 < t < 1$ and hence $0 < \cos t < \cos t^2$. In particular, we have $2 \sin t \cos t < 2t \cos t^2$. Integrating from 0 to x, we obtain the claimed inequality.

Problem 4.8. Let $P(x) = a_0 + a_1 x + \cdots + a_n x^n$ be a real polynomial of degree $n \geq 2$ such that

$$0 < a_0 < -\sum_{j=1}^{\left[\frac{n}{2}\right]} \frac{1}{2k + 1} a_{2k}.$$

Show that $P(x)$ has a real zero in $(-1, 1)$.

Solution 4.8. The hypothesis implies that

$$\int_{-1}^{1} P(x)\, dx = \sum_{j=0}^{\left[\frac{n}{2}\right]} \frac{1}{2k + 1} a_{2k} < 0.$$

Thus there exists $x \in (-1, 1)$ with $P(x) < 0$. Since $P(0) = a_0 > 0$, we can apply the intermediate value theorem to find that P has a real zero in $(-1, 1)$.

Problem 4.9. Find the minimum of β and the maximum of α for which

$$\left(1 + \frac{1}{n}\right)^{n+\alpha} \leq e \leq \left(1 + \frac{1}{n}\right)^{n+\beta}$$

holds for all $n \in \mathbb{Z}_+$.

Solution 4.9. Considering the logarithms, we have

$$\alpha_{max} = \inf_n \left\{ \frac{1}{\log(1 + 1/n)} - n \right\},$$

$$\beta_{min} = \sup_n \left\{ \frac{1}{\log(1 + 1/n) - n} \right\}.$$

The function $F(x) = \frac{1}{\log(1+1/x)} - x$ is monotonically increasing for positive x, since the derivative $F'(x)$ is positive. Therefore

$$\alpha_{max} = \frac{1}{\log 2} - 1$$

and

$$\beta_{min} = \lim_{n \to \infty} F(n).$$

Using the Maclaurin series for the function $\log(1 + 1/x)$, we obtain

$$F(n) = \frac{1}{\frac{1}{n} - \frac{1}{2n^2} + \mathcal{O}(\frac{1}{n^3})} - n,$$

whence $\lim_{n \to \infty} F(n) = \frac{1}{2} = \beta_{min}$.

Problem 4.10. Prove that for nonnegative x, y, and z such that $x + y + z = 1$, the following inequality holds:

$$0 \le xy + yz + zx - 2xyz \le \frac{7}{27}.$$

Solution 4.10. We have $x, y, z \in [0, 1]$ and thus $0 \le xyz \le 1$, which turns into $xy + yz + zx - 2xyz \ge 3\sqrt[3]{x^2 y^2 z^2} - 2xyz \ge 3xyz - 2xyz = xyz \ge 0$. This yields the first inequality.

Also, if one of the three variables x, y, z is at least $\frac{1}{2}$, then $(1 - 2x)(1 - 2y)(1 - 2z) \le 0 < \frac{1}{27}$. Observe that it is not possible for two among the three variables to be greater than $\frac{1}{2}$, because this would imply that the third is negative. Further, if $x, y, z \le \frac{1}{2}$, then, by the means inequality, $(1 - 2x)(1 - 2y)(1 - 2z) \le \frac{1}{27}$. This implies that

$$1 - 2(x + y + z) + 4(xy + yz + zx) - 8xyz \le \frac{1}{27}$$

and thus $xy + yz + zx - 2xyz \le \frac{7}{27}$.

Second Solution:

Let $f(x, y, z) = xy + yz + zx - 2xyz$ and the associated Lagrange multiplier

$$F(x, y, z, \lambda) = f - \lambda(x + y + z - 1).$$

The critical points of F are given by the formula

$$\frac{\partial F}{\partial x} = \frac{\partial F}{\partial y} = \frac{\partial F}{\partial z} = \frac{\partial F}{\partial \lambda} = 0.$$

Therefore, we obtain the system of equations

$$x + y - 2xy - \lambda = 0,$$
$$x + z - 2xz - \lambda = 0,$$
$$y + z - 2yz - \lambda = 0,$$
$$x + y + z - 1 = 0.$$

If one variable, say z, is not equal to $\frac{1}{2}$, then $x = \frac{z-\lambda}{1-2z} = y$. If $x = y \neq \frac{1}{2}$, then $x = z = \frac{y-\lambda}{1-2y}$. Thus $x = y = z = \frac{1}{3}$.

If $x = y = \frac{1}{2}$, then $z = 0$. Therefore, we have the solutions $\left(\frac{1}{3}, \frac{1}{3}, \frac{1}{3}\right)$, $\left(\frac{1}{2}, \frac{1}{2}, 0\right)$ and their circular permutations.

The Jacobian matrix reads

$$J(x, y, z) = \det \begin{pmatrix} \frac{\partial^2 F}{\partial x^2} & \frac{\partial^2 F}{\partial x \partial y} & \frac{\partial^2 F}{\partial x \partial z} \\ \frac{\partial^2 F}{\partial y \partial x} & \frac{\partial^2 F}{\partial y^2} & \frac{\partial^2 F}{\partial y \partial z} \\ \frac{\partial^2 F}{\partial z \partial x} & \frac{\partial^2 F}{\partial z \partial y} & \frac{\partial^2 F}{\partial z^2} \end{pmatrix} = \det \begin{pmatrix} 0 & 1-2z & 1-2x \\ 1-2z & 0 & 1-2y \\ 1-2y & 1-2x & 0 \end{pmatrix},$$

and one computes

$$J\left(\frac{1}{3}, \frac{1}{3}, \frac{1}{3}\right) = \frac{2}{27} > 0, \qquad J\left(\frac{1}{2}, \frac{1}{2}, 0\right) = 0.$$

Now $f\left(\frac{1}{2}, \frac{1}{2}, 0\right) = \frac{1}{4} < f\left(\frac{1}{3}, \frac{1}{3}, \frac{1}{3}\right) = \frac{7}{27}$, and so the only extremal point is a maximum at $\left(\frac{1}{3}, \frac{1}{3}, \frac{1}{3}\right)$.

Problem 4.11. Consider the sequence of nonzero complex numbers $a_1, \ldots, a_n, \ldots$ with the property that $|a_r - a_s| > 1$ for $r \neq s$. Prove that

$$\sum_{n=1}^{\infty} \frac{1}{a_n^3}$$

converges.

Solution 4.11. Let $S_k = \{n \in \mathbb{Z}_+; k < |a_n| \leq k+1\}$. The disks $D_n = \{z; |z - a_n| \leq 1/2\}$ are pairwise disjoint, according to the hypothesis. If one considers those disks D_n for which $n \in S_k$, then these disks are contained in the annulus

$$C_k = \left\{z \in \mathbb{C}; k - \frac{1}{2} \leq |z| \leq k + \frac{3}{2}\right\}.$$

Thus the total area covered by these disks is bounded from above by the area of the annulus, i.e.,

$$\operatorname{card}(S_k)\frac{\pi}{4} \le \pi\left(\left(k+\frac{3}{2}\right)^2 - \left(k-\frac{1}{2}\right)^2\right) = 2\pi(k+1),$$

and thus $\operatorname{card}(S_k) \le 8(2k+1)$ if $k > 0$. In the same way, for $k = 0$, we obtain $\operatorname{card}(S_0) \le 9$:

$$\sum_{n \in S_k} \frac{1}{|a_n|^3} \le \frac{\operatorname{card}(S_k)}{k^3} \le \frac{8(2k+1)}{k^3} \le \frac{24}{k^2},$$

$$\sum_{n=1}^{\infty} \frac{1}{|a_n|^3} = \sum_{k=0}^{\infty}\sum_{n \in S_k} \frac{1}{|a_n|^3} \le \sum_{n \in S_0} \frac{1}{|a_n|^3} + \sum_{k=1}^{\infty} \frac{24}{k^2} < \infty,$$

and the series converges absolutely.

Problem 4.12. Consider the sequence S_n given by

$$S_n = \frac{n+1}{2^{n+1}} \sum_{i=1}^{n} \frac{2^i}{i}.$$

Find $\lim_{n \to \infty} S_n$.

Solution 4.12. We have

$$S_{n+1} = \frac{n+2}{2^{n+2}} \sum_{i=1}^{n+1} \frac{2^i}{i} = \frac{n+2}{2(n+1)}(S_n + 1),$$

so that

$$S_{n+2} - S_{n+1} = \frac{(n+2)^2(S_{n+1} - S_n) - S_{n+1} - 1}{2(n+1)(n+2)}.$$

Since $S_n \ge 0$, we obtain $S_{n+2} - S_{n+1} \le 0$, and the sequence S_n is bounded decreasing, thus convergent. If we set s for its limit, then s satisfies

$$s = \lim_{n \to \infty} \frac{n+2}{2(n+1)}(S_n + 1) = \frac{1}{2}(s+1)$$

and thus $s = 1$.

Comments 80 *By integrating the geometric series formula, we obtain*

$$\lim_{n \to \infty} \frac{n}{a^n} \sum_{i=1}^{n} \frac{a^{i-1}}{i} = \frac{1}{a-1}.$$

For $a = 2$, we recover the previous problem.

Problem 4.13. Prove that whenever $a, b > 0$, we have

$$\int_0^1 \frac{t^{a-1}}{1+t^b}\,dt = \frac{1}{a} - \frac{1}{a+b} + \frac{1}{a+2b} - \frac{1}{a+3b} + \cdots.$$

Solution 4.13. Consider the identity

$$\frac{t^{a-1}}{1+t^b} = t^{a-1}\left(\sum_{k=0}^{n-1}(-1)^k t^{kb} + \frac{(-1)^n t^{bn}}{1+t^b}\right).$$

Integrating between 0 and 1 and using the fact that

$$\lim_{n\to\infty}\left|\int_0^1 \frac{t^{bn}}{1+t^b}\,dt\right| = 0,$$

we obtain the equality from the statement.

Problem 4.14. Let $a, b, c, d \in \mathbb{Z}_+^*$, and $r = 1 - \frac{a}{b} - \frac{c}{d}$. If $r > 0$ and $a + c \leq 1982$, then $r > \frac{1}{1983^3}$.

Solution 4.14. We have three cases to analyze:

1. If $b, d \geq 1983$, then $r \geq 1 - \frac{a+c}{1983} \geq 1 - \frac{1982}{1983} > \frac{1}{1983^3}$.

2. If $b, d \leq 1983$, then $r = \frac{bd-ad-bc}{bd} > 0$, so that $bd - ad - bc \geq 1$. Thus $r \geq 1/bd > 1/1983^3$.

3. Suppose now that $b < 1983 < d$. If $d > 1983^2$ and $r < \frac{1}{1983^3}$, then $\frac{c}{d} < \frac{1982}{1983^2} < \frac{1}{1983}$; thus $1 - \frac{a}{b} < \frac{1}{1983}$. This implies that $b > 1983(b-a) > 1983$ because $a \leq b + 1$, which is absurd. If $d \leq 1983^2$, then, as in the case 2 above, we have $r \geq \frac{1}{bd} > \frac{1}{1983^3}$.

Problem 4.15. Let $a_i \in R$. Prove that

$$n\min(a_i) \leq \sum_{i=1}^n a_i - S \leq \sum_{i=1}^n a_i + S \leq n\max(a_i),$$

where $(n-1)S^2 = \sum_{1\leq i<j\leq n}(a_i - a_j)^2$, with $S \neq 0$, and with equality if and only if $a_1 = \cdots = a_n$.

Solution 4.15. Let us assume $a_1 \leq a_2 \leq a_3 \leq \cdots \leq a_n$. Then

$$S^2 = \frac{1}{n-1}\sum_{i=2}^n\sum_{j=1}^{i-1}(a_i - a_j)^2$$

$$\leq \frac{1}{n-1}\sum_{i=2}^n(i-1)(a_i - a_1)^2 \leq \sum_{i=2}^n(a_i - a_1)^2 \leq \left(\sum_{i=1}^n(a_i - a_1)\right)^2.$$

Extracting the radicals, we obtain $na_1 \leq \sum_{i=1}^n a_i - S$. Similarly,

$$S^2 \leq \frac{1}{n-1}\sum_{j=1}^{n-1}(n-j)(a_n - a_j)^2 \leq \left(\sum_{j=1}^{n-1}(a_n - a_j)\right)^2,$$

and thus $\sum_{k=1}^n a_k + S \leq na_n$. It is now clear that equality above implies $S = 0$, and therefore $a_1 = a_2 = \cdots = a_n$.

Problem 4.16. Consider a periodic function $f : \mathbb{R} \to \mathbb{R}$, of period 1 that is nonnegative, concave on $(0, 1)$ and continuous at 0. Prove that for all real numbers x and natural numbers n the following inequality $f(nx) \le nf(x)$ is satisfied.

Solution 4.16. It is sufficient to discuss the case $0 < x < 1$. Since f is concave, we have $\frac{f(\xi)}{\xi} < \frac{f(x)}{x}$ for any ξ with $x < \xi < 1$. Thus $f(\xi) < \frac{\xi}{x} f(x)$. If $0 < \xi < x$, then similarly $f(\xi) < \frac{1-\xi}{1-x} f(x)$. Let $\xi = \{nx\}$, where the brackets $\{a\}$ denote the fractional part $a - [a]$.

If $\xi > x$, then we apply the first inequality above, and use the periodicity and the fact that $\frac{\xi}{x} \le n$ in order to get the claim.

If $\xi \le x$, then we have two cases:
1. If $x \le 1 - \frac{1}{n}$, then $\frac{1-\xi}{1-x} \le \frac{1}{1/n} = n$, and we apply the second inequality above.
2. If $x = 1 - \frac{1}{n} + \alpha$ for some $0 < \alpha < \frac{1}{n}$, then $nx = n-1+n\alpha$, and so $\{nx\} = n\alpha$.

Since $\frac{1-\xi}{1-x} = \frac{1-n\alpha}{\frac{1}{n} - \alpha} = n$, we use the second inequality again and the claim follows.

Comments 81 *When $f(x) = |\sin x|$, we obtain the well-known inequality $|\sin nx| \le n|\sin x|$.*

Problem 4.17. Let (a_n) be a sequence of positive numbers such that

$$\lim_{n\to\infty} \frac{(a_1 + \cdots + a_n)}{n} < \infty, \quad \lim_{n\to\infty} \frac{a_n}{n} = 0.$$

Does this imply that $\lim_{n\to\infty} \frac{a_1^2 + \cdots + a_n^2}{n^2} = 0$?

Solution 4.17. The answer is yes. If the sequence a_n is bounded, it is obvious. If the sequence is not bounded, consider the subsequence a_{n_k} defined by the properties that n_1 is the smallest integer such that $a_{n_1} > a_1$, and n_k the smallest integer $n_k > n_{k-1}$ such that $a_{n_k} > a_{n_{k-1}}$.

If $n_k \le n < n_{k+1}$, then we have

$$0 \le \frac{a_1^2 + \cdots + a_n^2}{n^2} < \left(\frac{a_1 + \cdots + a_n}{n} \right) \frac{a_{n_k}}{n_k}.$$

Passing to the limit, we find that the claim holds.

Problem 4.18. Show that for a fixed $m \ge 2$, the following series converges for a single value of x:

$$S(x) = 1 + \ldots + \frac{1}{m-1} - \frac{x}{m} + \frac{1}{m+1} + \cdots + \frac{1}{2m-1} - \frac{x}{2m}$$
$$+ \frac{1}{2m+1} + \cdots + \frac{1}{3m-1} - \frac{x}{3m} + \cdots .$$

Solution 4.18. Let us denote by $S_n(x) = \sum_{k=1}^{n}(\frac{1}{(k-1)m+1} + \cdots + \frac{1}{km-1} - \frac{x}{km})$ the truncation of $S(x)$. Assume that $S(x)$ and $S(y)$ are finite for two distinct values x and y.

We now observe that $S_n(x) - S_n(y) = \frac{y-x}{m} \sum_{k=1}^{n} \frac{1}{k}$, and therefore

$$\lim_{n \to \infty} |S_n(x) - S_n(y)| = \infty,$$

which contradicts our assumptions.

Comments 82 *If x_m is the value for which $S(x)$ converges, then $S(x_m) = \log m$.*

Problem 4.19. Prove that if $z_i \in \mathbb{C}, 0 < |z_i| \le 1$, then $|z_1 - z_2| \le |\log z_1 - \log z_2|$.

Solution 4.19. From Lagrange's mean value theorem, we have $|\log r_1 - \log r_2| = \frac{1}{\xi}|r_1 + r_2|$ for any $0 < r_1, r_2 < 1$ and $\xi \in [0, 1]$. Thus $|\log r_1 - \log r_2| \le |r_1 - r_2|$. Write $z_j = r_j \exp(i\theta_j)$. We have

$$(z_1 - z_2)^2 = r_1^2 + r_2^2 - 2r_1 r_2 \cos(\theta_1 - \theta_2) = (r_1 - r_2)^2 + 2r_1 r_2(1 - \cos(\theta_1 - \theta_2))$$

$$= (r_1 - r_2)^2 + 4r_1 r_2 \sin^2 \frac{1}{2}(\theta_1 - \theta_2) \le (r_1 - r_2)^2 + r_1 r_2(\theta_1 - \theta_2)^2$$

$$\le (r_1 - r_2)^2 + (\theta_1 - \theta_2)^2 \le (\log r_1 - \log r_2)^2 + (\theta_1 - \theta_2)^2$$

$$= |\log z_1 - \log z_2|^2.$$

We have used above the inequality $\sin \theta < |\theta|$, and its consequence $4\sin^2 \frac{1}{2}(\theta_1 - \theta_2) \le (\theta_1 - \theta_2)^2$.

Problem 4.20. The roots of the function of a complex variable

$$\zeta_4(s) = 1 + 2^s + 3^s + 4^s$$

are simple.

Solution 4.20. Suppose that $\zeta_4(s)$ and $\zeta_4'(s)$ have common zeros. Let us set $x = 2^s$. Then we have

$$1 + x + x^2 = -3^s \quad \text{and} \quad x \log 2 + x^2 \log 4 = -3^s \log 3,$$

so that

$$x^2(\log 3/2) + x(\log 3/2) + \log 3 = 0.$$

The discriminant of this quadratic equation is positive; hence the roots are real: $2^s = x \in \mathbb{R}, 3^s = -(1 + x + x^2) \in \mathbb{R}$.

It is known that s cannot be real, since ζ_4 has no real zeros. However, the reality condition above shows that the (nonzero) imaginary part of s must be an integer multiple of both $\frac{\pi i}{\log 2}$ and $\frac{\pi i}{\log 3}$. This is impossible, because $\log 3/\log 2 \notin \mathbb{Q}$.

Comments 83 *A related function that generalizes the finite sum $\zeta_4(s)$ is the Riemann zeta function ζ defined by*

$$\zeta(s) = \sum_{k \ge 1} k^{-s}.$$

It is not hard to see that $\zeta(s)$ is defined for all complex s with real part $\Re(s) > 1$ and can be continued analytically in s. The function $\zeta(s)$ has zeros at the negative even integers $-2, -4, \ldots$ and one usually refers to them as the trivial zeros.

One of the most important problems in mathematics today, known as the Riemann hypothesis, is to decide whether all nontrivial zeros of $\zeta(s)$ in the complex plane lie on the "critical line" consisting of points of real part $\frac{1}{2}$. The nontrivial zeros correspond precisely to the zeros of the following even entire function:

$$\xi(t) = \frac{\pi^{-s/2}s(s-1)}{2}\Gamma\left(\frac{s}{2}\right)\zeta(s),$$

where $s = \frac{1}{2} + \sqrt{-1}t$ and Γ is the usual gamma function. This is known to hold true for the first 10^{13} zeros and for 99 percent of the zeros. The validity of the conjecture is equivalent to saying that

$$\pi(n) - \int_0^n \frac{dt}{\log t} = \mathcal{O}(\sqrt{n}\log n),$$

where $\pi(n)$ counts the number of primes less than n.

The Riemann hypothesis was included as one of the seven Millennium Problems (for which the Clay Mathematics Institute will award one million dollar for a solution). The interested reader might consult:

- B. Conrey: *The Riemann hypothesis,* Notices Amer. Math. Soc. March, 2003, 341-353.
- E. Bombieri: *Problems of the millennium: The Riemann Hypothesis,* The Clay Mathematics Institute, 2000.
  http://www.claymath.org/millennium/Riemann_Hypothesis/

Problem 4.21. Compute

$$I = \int_0^{\pi/2} \log \sin x \, dx.$$

Solution 4.21. We have

$$I = \int_0^{\pi/2} \log \cos x \, dx = \int_{\pi/2}^{\pi} \log \sin x \, dx = \frac{1}{2}\int_0^{\pi} \log \sin x \, dx$$

and thus

$$2I = 2\int_0^{\pi/2} \log \sin 2x \, dx$$

$$= 2\left(\int_0^{\pi/2} \log 2 \, dx + \int_0^{\pi/2} \log \sin x \, dx + \int_0^{\pi/2} \log \cos x \, dx\right) = 4I + \pi\log 2$$

and thus $I = -\frac{\pi}{2}\log 2$.

Problem 4.22. Let f a be positive, monotonically decreasing function on $[0, 1]$. Prove the following inequality:

$$\frac{\int_0^1 x f^2(x)\,dx}{\int_0^1 x f(x)\,dx} \le \frac{\int_0^1 f^2(x)\,dx}{\int_0^1 f(x)\,dx}.$$

Solution 4.22. The above inequality is equivalent to

$$\int_0^1 f^2(x)\,dx \int_0^1 y f(f)\,dy - \int_0^1 y f^2(y)\,dy \int_0^1 f(x)\,dx \ge 0,$$

which can be rewritten as follows:

$$I = \int_0^1 \int_0^1 f(x) f(y) y (f(x) - f(y))\,dx\,dy \ge 0.$$

Interchanging the variables and the integration order, we obtain

$$2I = \int_0^1 \int_0^1 f(x) f(y)(y - x)(f(x) - f(y))\,dx\,dy \ge 0.$$

Since f is decreasing, we have $(y - x)(f(x) - f(y)) \ge 0$ and thus $2I \ge 0$, as claimed.

Comments 84 *It can be proven by the same method that*

$$\int_0^1 f(x) g(x) w(x)\,dx \int_0^1 w(x)\,dx \le \int_0^1 f(x) w(x)\,dx \int_0^1 g(x) w(x)\,dx,$$

where f is decreasing, g is increasing, and w is nonnegative, with strict inequality if f and g are not constant on the support of w.

Problem 4.23. Prove that every real number x such that $0 < x \le 1$ can be represented as an infinite sum

$$x = \sum_{k=1}^{\infty} \frac{1}{n_k},$$

where the n_k are natural numbers such that $\frac{n_{k+1}}{n_k} \in \{2, 3, 4\}$.

Solution 4.23. We have two cases to consider, $\frac{1}{3} \le x \le 1$ and $0 < x < \frac{1}{3}$.

In the first case, let us put $n_0 = 1$ and $x_0 = x$. We define the sequence $x_k = \frac{n_k}{n_{k-1}} x_{k-1} - 1$, where

$$n_k = \begin{cases} 4n_{k-1}, & \text{if } \frac{1}{3} \le x_{k-1} < \frac{1}{2}, \\ 3n_{k-1}, & \text{if } \frac{1}{2} \le x_{k-1} < \frac{2}{3}, \\ 2n_{k-1}, & \text{if } \frac{2}{3} \le x_{k-1} < 1. \end{cases}$$

If $\frac{1}{3} \leq x_{k-1} \leq 1$, then it is immediate that $\frac{1}{3} \leq x_k \leq 1$ and so, $\frac{1}{3} \leq x_k \leq 1$. Further, the sequence $\{n_k\}$ is unbounded, which implies that

$$\lim_{k \to \infty} \frac{x_k}{n_k} = 0.$$

The recurrence relation reads

$$\frac{x_k}{n_k} = \frac{x_{k-1}}{n_{k-1}} - \frac{1}{n_k},$$

and hence

$$\frac{x_k}{n_k} = \frac{x_0}{n_0} - \sum_{i=1}^{k} \frac{1}{n_i} = x - \sum_{i=1}^{k} \frac{1}{n_i}.$$

Thus by letting k go to infinity, $x = \sum_{i=1}^{\infty} \frac{1}{n_i}$.

In the second case, $0 < x < \frac{1}{3}$, we choose a natural number $k \geq 4$ such that $1/k \leq x < 1/(k-1)$. Set $N = 2(k-1)$. Then the number $x_1 = Nx - 1$ satisfies $\frac{1}{3} < x_1 < 1$. From the above,

$$x_1 = \sum_{k=1}^{\infty} \frac{1}{n_k},$$

where $\frac{n_{k+1}}{n_k} \in \{2, 3, 4\}$. Then we write

$$x = \frac{1}{N} + \frac{x_1}{N} = \frac{1}{N} + \sum_{k=1}^{\infty} \frac{1}{Nn_k} = \sum_{k=1}^{\infty} \frac{1}{m_k},$$

where $m_k = Nn_{k-1}$, for $k \geq 2$, and $m_1 = N$.

Comments 85 *In general, the representation is not unique, since for example,*

$$\frac{1}{2} = \sum_{k=2}^{\infty} \frac{1}{3^k} = \sum_{k=1}^{\infty} \frac{1}{2^k}.$$

Moreover, the result of the problem holds as well for those x that belong to the interval $[\frac{4}{3}, 2]$.

Problem 4.24. Let S_n denote the set of polynomials $p(z) = z^n + a_{n-1}z^{n-1} + \cdots + a_1 z + 1$, with $a_i \in \mathbb{C}$. Find

$$M_n = \min_{p \in S_n} \left(\max_{|z|=1} |p(z)| \right).$$

Solution 4.24. Let us show that $M_n = 2$ if $n \geq 2$. We have first $M_n \leq \max_{|z|=1} |z^n + 1| = 2$.

Suppose now that $M_n < 2$, for some n, so that there exists a polynomial $p(z) = z^n + a_{n-1}z^{n-1} + \cdots + a_1 z + 1$ with the property that $\max_{|z|=1} |p(z)| < 2$. This implies that

$$-2 < \Re(p(z)) < 2, \text{ for } z \in \mathbb{C}, \text{ with } |z| = 1.$$

Let ξ_j, $j = 1, \ldots, n$ be the nth roots of unity. Sum the inequalities above for $z = \xi_j$, and get

$$-2n < \Re \sum_{j=1}^{n} p(\xi_j) < 2n.$$

On the other hand, an immediate computation shows that

$$\sum_{j=1}^{n} p(\xi_j) = 2n$$

for any p as above, which is a contradiction. Thus our claim follows.

Problem 4.25. Find an explicit formula for the value of F_n that is given by the recurrence $F_n = (n + 2)F_{n-1} - (n - 1)F_{n-2}$ and initial conditions $F_1 = a$, $F_2 = b$.

Solution 4.25. We will use the so-called Laplace method. Let $\vartheta : \mathbb{R} \to \mathbb{R}$ be a smooth function such that

$$F_n = \int_{\alpha}^{\beta} t^{n-1} \vartheta(t) \, dt.$$

The recurrence satisfied by F_n is then a consequence of the following differential equation:

$$\frac{d\vartheta}{\vartheta} = \frac{1 - 3t + t^2}{t - t^2} \, dt.$$

This differential equation can be integrated, and one obtains the solution $\vartheta = e^{-t} t(t - 1)$. Further, the integral's limits α and β are among the roots of the equation

$$t^{n+1}(t - 1)e^{-t} = 0,$$

yielding $t_1 = 0$, $t_2 = \infty$, $t_3 = 1$. The general solution is therefore

$$F_n = C_1 \int_0^{\infty} t^n (t - 1)e^{-t} \, dt + C_2 \int_0^1 t^n (t - 1) e^{-t} dt$$

$$= C_1 n!n + C_2 \left(n!n \left(1 - \frac{1}{e} \sum_{k=0}^{n} \frac{1}{k!} \right) - \frac{1}{e} \right)$$

$$= (3b - 11a)n!n + (4a - b) \left(1 + n!n \sum_{k=0}^{n} \frac{1}{k!} \right).$$

Problem 4.26. Find the positive functions $f(x, y)$ and $g(x, y)$ satisfying the following inequalities:

$$\left(\sum_{i=1}^{n} a_i b_i \right)^2 \leq \left(\sum_{i=1}^{n} f(a_i, b_i) \right) \left(\sum_{i=1}^{n} g(a_i, b_i) \right) \leq \left(\sum_{i=1}^{n} a_i^2 \right) \left(\sum_{i=1}^{n} b_i^2 \right)$$

for all $a_i, b_i \in \mathbb{R}$ and $n \in \mathbb{Z}_+$.

Solution 4.26.

$$(ab)^2 \le f(a, b)g(a, b) \le a^2b^2$$

and thus

$$(1) \qquad f(a, b)g(a, b) = a^2b^2.$$

For $n = 2$, use the hypothesis

$$2a_1b_1a_2b_2 \le f(a_1, b_1)g(a_2, b_2) + f(a_2, b_2)g(a_1, b_1) \le a_1^2b_2^2 + a_2^2b_1^2$$

and the identity (1) to find that

$$(2) \qquad 2 \le \frac{f(a_1, b_1)}{f(a_2, b_2)}\frac{a_2b_2}{a_1b_1} + \frac{f(a_2, b_2)}{f(a_1, b_1)}\frac{a_1b_1}{a_2b_2} \le \frac{a_1b_2}{a_2b_1} + \frac{a_2b_1}{a_1b_2}.$$

We replace $(a_1, b_1) = (a, b)$ and $(a_2, b_2) = (\lambda a, \lambda b)$; then

$$2 \le \frac{f(a, b)}{f(\lambda a, \lambda b)}\lambda^2 + \frac{f(\lambda a, \lambda b)}{f(a, b)}\lambda^{-2} \le 2,$$

from which we derive

$$f(\lambda a, \lambda b) = \lambda^2 f(a, b).$$

Set $f(a) = f(a, 1)$, and so $f(a, b) = b^2 f(a/b)$. Then (2) becomes

$$(3) \qquad 2 \le \frac{f(a)/a}{f(b)/b} + \frac{f(b)/b}{f(a)/a} \le \frac{a}{b} + \frac{b}{a}.$$

This implies that when $a \ge b$, we have $\frac{f(a)b}{f(b)a} \le \frac{a}{b}$ and $\frac{f(b)a}{f(a)b} \le \frac{a}{b}$. This yields

$$(4) \qquad f(b) \le f(a), \quad \frac{f(a)}{a^2} \le \frac{f(b)}{b^2}, \quad (a \ge b).$$

Conversely, if f, g satisfy (1) and (4), then by applying the inequality (2) for all pairs of indices $i \ne j \le n$, we find that

$$2\sum_{i,j} a_ib_ia_jb_j \le \sum_{i,j}[f(a_i, b_i)g(a_j, b_j) + f(a_j, b_j)g(a_i, b_i)] \le \sum_{i,j}(a_i^2b_j^2 + a_j^2b_i^2),$$

and the inequality from the statement follows.

Comments 86 *If $f(x, y) = x^2 + y^2$, $g(x, y) = x^2y^2/(x^2 + y^2)$, then we obtain an inequality due to Hardy, Littlewood, and Pólya. By choosing $f(x, y) = x^{1+\alpha}y^{1-\alpha}$, $g(x, y) = x^{1-\alpha}y^{1+\alpha}$, we obtain the Calderbank inequality.*

Problem 4.27. Let $g \in C^1(\mathbb{R})$ be a smooth function such that $g(0) = 0$ and $|g'(x)| \le |g(x)|$. Prove that $g(x) = 0$.

Solution 4.27. Let $x \geq 0$. We have

$$|g(x)| \leq \left| \int_0^x g'(t) \, dt \right| \leq \int_0^x |g'(t)| \, dt \leq \int_0^x |g(t)| \, dt.$$

Let $a > 0$ be arbitrary. Since $g \in C^1(\mathbb{R})$, g is bounded on any bounded interval and thus there exists k, such that $|g(x)| \leq k$, for $0 \leq x \leq a$.

If $0 \leq t \leq x \leq a$, then $|g(t)| \leq k$, and so

$$|g(x)| \leq \int_0^x k \, dt = kx$$

for every x with $0 \leq x \leq a$. Now the condition $|g(t)| \leq kt$, for $0 \leq t \leq x \leq a$, implies that

$$|g(x)| \leq \int_0^x kt \, dt = \frac{1}{2}kx^2, \quad \text{for } 0 \leq x \leq a.$$

Continuing this way, we prove by induction that

$$|g(x)| \leq \frac{1}{n!}kx^n, \quad \text{provided that } 0 \leq x \leq a.$$

Putting $x = a$ and passing to the limit $n \to \infty$, we obtain

$$|g(a)| \leq \lim_{n \to \infty} \frac{1}{n!}ka^n = 0.$$

Since a was arbitrary, we have $g(x) = 0$ for all $x \in [0, \infty)$. A similar argument settles the case $x \in (-\infty, 0]$.

Problem 4.28. Let $a_1, b_1, c_1 \in \mathbb{R}_+$ such that $a_1 + b_1 + c_1 = 1$ and define

$$a_{n+1} = a_n^2 + 2b_nc_n, \quad b_{n+1} = b_n^2 + 2a_nc_n, \quad c_{n+1} = c_n^2 + 2a_nb_n.$$

Prove that the sequences (a_n), (b_n), (c_n) have the same limit and find that limit.

Solution 4.28. Let

$$u_n = a_n + b_n + c_n, \quad v_n = a_n + \omega b_n + \omega^2 c_n, \quad w_n = a_n + \omega^2 b_n + \omega c_n,$$

where $\omega^3 = 1$ is a third root of unity. The given recurrence for a_n, b_n, c_n induces the following recurrence relations:

$$u_{n+1} = u_n^2 = 1, \quad v_{n+1} = w_n^2, \quad w_{n+1} = v_n^2.$$

Thus w_{n+1} equals either v_1^{2n} or w_1^{2n}, depending on the parity of n (and similarly for v_{n+1}). Since v_1 and w_1 are convex combinations with positive coefficients of the complex points $1, \omega, \omega^2$, it follows that $|v_1|, |w_1| < 1$. But then we derive that $\lim_{n \to \infty} v_n = \lim_{n \to \infty} w_n = 0$. Since u_n is constant, we obtain

$$\lim_{n\to\infty} a_n = \lim_{n\to\infty} \frac{1}{3}(u_n + v_n + w_n) = \frac{1}{3},$$

$$\lim_{n\to\infty} b_n = \lim_{n\to\infty} \frac{1}{3}(u_n + \omega^2 v_n + \omega w_n) = \frac{1}{3},$$

$$\lim_{n\to\infty} c_n = \lim_{n\to\infty} \frac{1}{3}(u_n + \omega v_n + \omega^2 w_n) = \frac{1}{3}.$$

Problem 4.29. Consider the sequence (a_n), given by $a_1 = 0$, $a_{2n+1} = a_{2n} = n - a_n$. Prove that $a_n = \frac{n}{3}$, for infinitely many values of n. Does there exist an n such that $|a_n - \frac{n}{3}| > 2005$? Also, prove that

$$\lim_{n\to} \frac{a_n}{n} = \frac{1}{3}.$$

Solution 4.29. 1. We have $a_n = \left[\frac{n}{2}\right] - a_{\left[\frac{n}{2}\right]}$, and thus

$$a_n = \left[\frac{n}{2}\right] - \left[\frac{n}{4}\right] + \left[\frac{n}{8}\right] - \left[\frac{n}{10}\right] + \cdots .$$

If $n = 3 \cdot 2^k$, then we have

$$a_{3\cdot 2^k} = \frac{3 \cdot 2^k}{2} - \frac{3 \cdot 2^k}{4} + \cdots + \frac{3 \cdot 2^k}{2^k} - \left[\frac{3}{2}\right]$$
$$= 3(2^{k-1} + (-2^{k-2}) + 2^{k-3} + (-2^{k-4}) \cdots) - 1$$
$$= 3 \cdot \frac{2^k + 1}{3} - 1 = 2^k,$$

and therefore $a_n = \frac{n}{3}$.

2. Set $n_k = \frac{4^k - 1}{3}$. Then n_k is odd and satisfies the recurrence $n_k - 1 = 4n_{k-1}$. This implies that

$$a_{n_k} = a_{n_k-1} = a_{4n_{k-1}} = 2n_{k-1} - a_{2n_{k-1}} = 2n_{k-1} - n_{k-1} + a_{n_{k-1}} = n_{k-1} + a_{n_{k-1}}$$

and thus

$$a_{n_k} = n_{k-1} + n_{k-2} + \cdots + n_1 + a_1 = \frac{1}{3}\left(\frac{4^k - 1}{3} - k\right).$$

Thus $\left|a_{n_k} - \frac{n_k}{3}\right| = \frac{k}{3}$, and so $\left|a_{n_k} - \frac{n_k}{3}\right|$ can be arbitrarily large.

3. Use the fact that $\frac{n}{2^k} - 1 \le \left[\frac{n}{2^k}\right] \le \frac{n}{2^k}$ and find that

$$\left(\sum_{k=1}^{[\log n]} (-1)^k \frac{n}{2^k}\right) - [\log n] \le a_n \le \left(\sum_{k=1}^{[\log n]} (-1)^k \frac{n}{2^k}\right) + [\log n] + 1,$$

and after dividing by n and letting n go to infinity, we obtain the claim.

Problem 4.30. Compute the integral

$$f(a) = \int_0^1 \frac{\log(x^2 - 2x \cos a + 1)}{x} \, dx.$$

Solution 4.30.

$$f\left(\frac{a}{2}\right) + f\left(\pi - \frac{a}{2}\right) = \int_0^1 \frac{\log(x^2 - 2x \cos(\frac{a}{2}) + 1)(x^2 + 2x \cos(\pi - \frac{a}{2}) + 1)}{x} \, dx$$

$$= \int_0^1 \frac{\log(x^4 - 2x^2 \cos a + 1)}{x} \, dx.$$

Make the variable change $x = \sqrt{t}$ and find that the last integral equals

$$\frac{1}{2} \int_0^1 \frac{(\log t^2 - 2t \cos a + 1)}{t} \, dt = \frac{f(a)}{2}, \quad \text{for } a \in [0, 2\pi].$$

Therefore, f satisfies the identity

$$f\left(\frac{a}{2}\right) + f\left(\pi - \frac{a}{2}\right) = \frac{f(a)}{2}.$$

By differentiating twice (we have f is C^2), we obtain

$$f''\left(\frac{a}{2}\right) + f''\left(\pi - \frac{a}{2}\right) = 2f''(a).$$

Suppose that $f''(a)$ has the maximum M and the minimum m, and let a_0 be such that

$$f''(a_0) = M, \quad a_0 \in [0, 2\pi].$$

Putting $a = a_0$ in the identity above we get

$$f''\left(\frac{a_0}{2}\right) + f''\left(\pi - \frac{a_0}{2}\right) = 2f''(a_0) = 2M,$$

from which we obtain $f''(a_0/2) = M$. Iterating this procedure, we find that $f''(a_0) = f''\left(\frac{a_0}{2^n}\right) = M$, for any M. From the continuity of f'', we find that

$$\lim_{n \to 0} f''\left(\frac{a_0}{2^n}\right) = f''(0) = M.$$

But now a similar argument dealing with the minimum shows that $f''(0) = m$ and therefore $M = m$; hence f'' is constant. This implies that $f(a) = \alpha a^2/2 + \beta a + \gamma$. Substituting in the identity above, we obtain the following identities in the coefficients:

$$-\pi\alpha = \beta, \quad \pi^2\alpha/2 + \beta\pi + 2\gamma = \gamma/2.$$

Using now $f'(\frac{\pi}{2}) = \frac{\pi}{2}$, it follows that $\pi\alpha/2 + \beta = \pi/2$, therefore $\alpha = -1$, $\beta = \pi$, $\gamma = -\pi^2/3$. Thus we obtain

$$f(a) = -\frac{a-2}{2} + \pi a - \frac{\pi^2}{3}.$$

Problem 4.31. Let $-1 < a_0 < 1$, and define $a_n = \left(\frac{1}{2}(1 + a_{n-1})\right)^{1/2}$ for $n \geq 1$. Find the limits A, B, and C of the sequences

$$A_n = 4^n(1 - a_n), \quad B_n = a_1 \cdots a_n, \quad C_n = 4^n(B - a_1 a_2 \cdots a_n).$$

Solution 4.31. Let $a_0 = \cos\theta$. By making use of the formula $\cos^2\theta/2 = (1 + \cos\theta)/2$ and a recurrence on n, we derive that $a_n = \cos(\theta/2^n)$. Then

$$A_n = 4^n(1 - a_n) = 4^n\left(\frac{\theta^2}{2 \cdot 4^n} - \frac{\theta^4}{24 \cdot 4^{2n}} + \mathcal{O}(4^{-3n})\right) = \frac{\theta^2}{2} - \frac{\theta^4}{24 \cdot 4^n} + \mathcal{O}(4^{-2n}).$$

Thus $A = \frac{\theta^2}{2}$. Further, $\sin(\theta/2^n)B_n = \cos(\theta/2)\cos(\theta/2^2)\cdots\cos(\theta/2^n)\sin(\theta/2^n)$ and thus

$$B_n = \frac{\sin\theta}{2^n\sin(\theta/2^n)} = \frac{\sin\theta}{\theta} + \frac{\sin\theta}{6 \cdot 4^n} + \mathcal{O}(4^{-2n})$$

and thus $B = \frac{\sin\theta}{\theta}$. Finally, $C_n = -\theta\sin\frac{\theta}{6} + \mathcal{O}(4^{-n})$, and so $C = -\theta\sin\frac{\theta}{6}$.

Comments 87 *The same method works when $a_0 > 1$, by making use of the hyperbolic trigonometric functions. If we put*

$$\theta = \log\left(a_0 + \sqrt{a_0^2 - 1}\right) = \text{arccosh } a_0,$$

then $A = -\theta^2/2$, $B = \frac{\sinh\theta}{6}$, $C = -6\sinh\frac{\theta}{6}$.

Problem 4.32. Consider the sequence given by the recurrence

$$a_1 = a, \quad a_n = a_{n-1}^2 - 2.$$

Determine those $a \in \mathbb{R}$ for which (a_n) is convergent.

Solution 4.32. For $|a| > 2$, we have $|a_n| > 2$ and consequently $a_{n+1} > a_n$ for all $n \geq 2$. Since the sequence is increasing, it is either convergent or tends to ∞. If we have a finite limit $a_n \to \lambda$, then it satisfies $\lambda = \lambda^2 - 2$ and thus $\lambda \in \{-1, 2\}$, contradicting our assumptions.

Therefore, we need $|a| \leq 2$ for the convergence. Let $\alpha \in \left(-\frac{\pi}{2}, \frac{\pi}{2}\right)$ be such that $2\cos\alpha = a$. By induction on n, we obtain

$$a_n = 2\cos 2^n\alpha.$$

Assume that α is not commensurable with π, that is, $\frac{\alpha}{\pi} \notin \mathbb{Q}$. Observe that the function $\varphi(x) = 2^x\alpha$ is continuous. Thus, the map $\varphi : \mathbb{R}/\pi\mathbb{Z} \to \mathbb{R}/\pi\mathbb{Z}$ sends the dense subset $\{n\alpha \pmod{\pi}, n \in \mathbb{Z}\}$ into a dense subset, which is $\{2^n\alpha \pmod{\pi}, n \in \mathbb{Z}\}$. Therefore, the image under $\cos$, namely $\{\cos 2^n\alpha; n \in \mathbb{Z}\}$, is dense in $[-1, 1]$, and thus the sequence (a_n) cannot be convergent.

If $\alpha/\pi = p/q$, where $p, q \in \mathbb{Z}$, then the sequence (a_n) is periodic for n large enough. Thus, in order for the sequence to be convergent it is necessary and sufficient that it be constant for n large enough. This condition reads

$$\cos(2^n\alpha) = \cos(2^{n+1}\alpha).$$

Thus

$$2^n\alpha = \pm 2^{n+1}\alpha + k\pi, \quad \text{where } k \in \mathbb{Z},$$

and so

$$\alpha \in \left\{\frac{m}{2^n 3}\pi, \text{ where } m, n \in \mathbb{Z}\right\}.$$

One sees that a_n is convergent and constant for $n \geq m$.

Problem 4.33. Let $0 < a < 1$ and $I = (0, a)$. Find all functions $f : I \to \mathbb{R}$ satisfying at least one of the conditions below:

1. f is continuous and $f(xy) = xf(y) + yf(x)$.
2. $f(xy) = xf(x) + yf(y)$.

Solution 4.33. 1. Let $g(x) = \frac{f(x)}{x}$. Then $g(\exp(x))$ is continuous and additive on I. This implies that $g(\exp(x))$ is linear and hence $g(x) = C \log x$, which implies $f(x) = Cx \log x$ for some constant C.

2. Give y the values x, x^2, x^3, and x^4. We find therefore that

$$f(x^2) = 2xf(x),$$
$$f(x^3) = xf(x) + x^2 f(x^2) = (x + 2x^3)f(x),$$
$$f(x^4) = xf(x) + x^3 f(x^3) = (x + x^4 + 2x^6)f(x),$$
$$f(x^4) = x^2 f(x) + x^2 f(x^2) = 4x^3 f(x).$$

From the last two lines, one derives that $f(x) = 0$ for all but those points x for which $x + x^4 + 2x^6 = 4x^3$. This means that f vanishes for all but finitely many (actually 6) values of x.

If $t \in I$ is such that $f(t) \neq 0$, then the first identity shows that $f(t^2) \neq 0$. By induction on n we obtain $f(t^{2^n}) \neq 0$. This means that f does not vanish for infinitely many values of the argument, which contradicts our former result. Thus f is identically zero.

Problem 4.34. If $a, b, c, d \in \mathbb{C}, ac \neq 0$, prove that

$$\frac{\max(|ac|, |ad + bc|, |bd|)}{\max(|a|, |b|)\max(|c|, |d|)} \geq \frac{-1 + \sqrt{5}}{2}.$$

Solution 4.34. Set $ab^{-1} = r$, $dc^{-1} = s$, $\frac{-1+\sqrt{5}}{2} = k$, $k^2 = 1 - k$, $0 < k < 1$. The inequality from the statement is equivalent to

$$\mu = \max(1, |r + s|, |rs|) \geq k \max(1, |r|)\max(1, |s|) = \nu.$$

(i) If $|r| \geq 1$, $|s| \geq 1$, then

$$\mu \geq |rs| > k|s||r| = \nu,$$

proving our claim.

(ii) If $|r| \le 1$, $|s| \ge 1$, then our inequality is equivalent to

$$\max(1, |r+s|, |rs|) \ge k|s|.$$

Moreover, if $k|s| \le 1$ or $|r+s| \ge k|s|$, then this is obviously true.

Further, let us assume that $k|s| > 1$ and $|r+s| < k|s|$. We have then $|r|+|r+s| \ge |s|$ and consequently,

$$|r| \ge |s| - |s+r| \ge |s| - k|s| = (1-k)|s| = k^2|s|.$$

Furthermore, $|rs| \ge k^2|s|^2 > k|s| > 1$; thus $\max(1, |r+s|, |rs|) \ge k|s|$ holds.

(iii) If $|r| \ge 1$, $|s| \le 1$, then the inequality is proved as in case (ii), by using the obvious symmetry.

Problem 4.35. Let $\sum_{i=1}^{\infty} x_i$ be a convergent series with decreasing terms $x_1 \ge x_2 \ge \cdots \ge x_n \ge \cdots > 0$ and let P be the set of numbers which can be written in the form $\sum_{i \in J} x_i$ for some subset $J \in \mathbb{Z}_+$. Prove that P is an interval if and only if

$$x_n \le \sum_{i=n+1}^{\infty} x_i \text{ for every } n \in \mathbb{Z}_+.$$

Solution 4.35. 1. Suppose that there exists p such that $x_p > \sum_{i=p+1}^{\infty} x_i$ and consider α such that $\sum_{i>p} x_i < \alpha < x_p$. Set $S(J) = \sum_{i \in J} x_i$. We claim that there does not exist any $J \subset \mathbb{Z}_+$ such that $S(J) = \alpha$. In fact, if $J \cap \{1, 2, \ldots, p\} \ne \emptyset$, then $\sum_{i \in J} x_i \ge x_p > \alpha$, because x_k is decreasing. On the other hand, if $J \cap \{1, 2, \ldots, p\} = \emptyset$, then $\sum_{i \in J} x_i \le \sum_{i>p} x_i < \alpha$. Finally, P contains numbers smaller than α, for instance $\sum_{i>p} x_i$, as well as numbers bigger than α, such as x_p. Thus P is not an interval.

2. Assume now that the hypothesis of the problem is satisfied and choose any element y from the interval $(0, S]$, where $S = \sum_{i=1}^{\infty} x_i$. We will show that there exists L such that $S(L) = y$.

Let n_1 be the smallest index such that $x_{n_1} < y$. There exists such a one because $\lim_{k \to \infty} x_k = 0$. By induction, we define n_k to be the smallest integer with the property that

$$x_{n_1} + \cdots + x_{n_k} < y.$$

Let $L = \{n_1, \ldots, n_k, \ldots\}$.

It is clear that $S(L) \le y$. If $p \in \mathbb{Z}_+ \setminus L$, then choose the smallest k with the property that $n_k > p$. Since p does not belong to L, we have

$$x_{n_1} + \cdots + x_{n_{k-1}} + x_p \ge y.$$

This implies that

$$\left(\sum_{i \in L}^{\infty} x_i \right) + x_p \ge y.$$

(i) Assume that the set $\mathbb{Z}_+ - L$ is finite. This set is empty only when $y = S$, which obviously belongs to P. Now take p to be the maximal element of $\mathbb{Z}_+ - L$. This implies that all elements bigger than p already belong to L and hence

$$L = \{n_1, \ldots, n_{k-1}\} \cup \{p+1, p+2, \ldots\}.$$

Then

$$\sum_{i \in L} x_i = \sum_{i=1}^{k-1} x_{n_1} + \sum_{j=p+1}^{\infty} x_j \geq \sum_{i=1}^{k-1} x_{n_i} + x_p \geq y,$$

from which we obtain that $\sum_{i \in L} x_i = y$.

(ii) If $\mathbb{Z}_+ - L$ is infinite, then for any $\varepsilon > 0$, one can choose some $p_\varepsilon \in \mathbb{Z}_+ - L$ such that $x_{p_\varepsilon} < \varepsilon$. The inequality above for p_ε implies that

$$\epsilon + \sum_{i \in L} x_i \geq y.$$

This is true for any positive ε, and thus $\sum_{i \in L} x_i = y$.

Problem 4.36. Does there exist a continuous function $f : (0, \infty) \to \mathbb{R}$ such that $f(x) = 0$ if and only if $f(2x) \neq 0$? What if we require only that f be continuous at infinitely many points?

Solution 4.36. 1. There exists some real number α such that $f(\alpha) = 0$, and thus $f(2\alpha) \neq 0$. Let $\gamma \in [\alpha, 2\alpha]$ be given by $\gamma = \sup\{x \in [\alpha, 2\alpha] \text{ such that } f(x) = 0\}$. Since f is continuous, we find that $f(\gamma) = 0$, and thus $\gamma < 2\alpha$.

Moreover, if $x_n \in \mathbb{R}$ is a sequence such that $\lim_n x_n = \gamma$, then $f(x_n) = 0$ for n large enough. In fact, otherwise there exists a subsequence $f(x_{n_k}) \neq 0$, which implies that $f(2x_{n_k}) = 0$, and thus $\lim_k f(2x_{n_k}) = 0 \neq f(2\gamma)$.

Finally, recall that γ is the supremum of those $x \in [\alpha, 2\alpha]$ for which $f(x) = 0$ and $\gamma < 2\alpha$. Thus for n large such that $\gamma + \frac{1}{n} < 2\alpha$, there exists x_n with $\gamma < x_n < \gamma + \frac{1}{n}$ for which $f(x_n) \neq 0$. This contradicts the previous claim. Thus there does not exist a continuous f as in the statement.

2. The function

$$f(x) = \begin{cases} 0, & \text{for } 2^{2k} < x \leq 2^{2k+1}, \text{ where } k \in \mathbb{Z}, \\ 1, & \text{for } 2^{2k-1} < x < 2^{2k}, \text{ where } k \in \mathbb{Z}, \end{cases}$$

satisfies the claim and has a countable number of discontinuities.

Problem 4.37. Find the smallest number a such that for every real polynomial $f(x)$ of degree two with the property that $|f(x)| \leq 1$ for all $x \in [0, 1]$, we have $|f'(1)| \leq a$. Find the analogous number b such that $|f'(0)| \leq b$.

Solution 4.37. Let $f(x) = ux^2 + vx + w$. We have

$$|f'(1)| = |2u + v| = |3f(1) - 4f(1/2) + f(0)| \leq 3|f(1)| + 4|f(1/2)| + |f(0)|$$
$$\leq 3 + 4 + 1 = 8.$$

Further, $a = 8$ because we have equality above for $f(x) = 8x^2 - 8x + 1$.

For the second case, we observe that

$$|f'(0)| = |v| = |4f(1/2) - 3f(0) - f(1)| \leq 8$$

and thus $b = 8$ by taking $f(x) = 8x^2 - 8x + 1$.

Problem 4.38. Let $f : \mathbb{R} \to \mathbb{R}$ be a function for which there exists some constant $M > 0$ satisfying

$$|f(x + y) - f(x) - f(y)| \leq M, \quad \text{for all } x, y \in \mathbb{R}.$$

Prove that there exists a unique additive function $g : \mathbb{R} \to \mathbb{R}$ such that

$$|f(x) - g(x)| \leq M, \quad \text{for all } x \in \mathbb{R}.$$

Moreover, if f is continuous, then g is linear.

Solution 4.38. Let us show first the existence of the upper limit $\varphi(x) = \limsup_{n \to \infty} \frac{f(nx)}{n}$.

In fact, using the hypothesis, one finds that $f(nx) \leq n(f(x) + M)$, and hence the sequence $\frac{f(nx)}{n}$ is bounded from above, and thus $\varphi(x)$ exists.

Furthermore,

$$\left| \frac{f(n(x+y))}{n} - \frac{f(nx)}{n} - \frac{f(ny)}{n} \right| \leq \frac{M}{n},$$

which yields $\varphi(x) + \varphi(y) = \varphi(x + y)$; thus φ is additive.

Next, observe that $f(nx) \leq n(f(x) + M)$ implies that $\left| \frac{nx}{n} - f(x) \right| \leq M$, and we can take $g(x) = \varphi(x)$.

If f is continuous, then it is easy to see that the lower limit, $\liminf_{n \to \infty} \frac{f(nx)}{n}$, coincides with $\varphi(x)$, and moreover, $\varphi(x)$ is continuous. An additive continuous function is therefore linear. In fact, we have $f(nx) = nf(x)$ for $n \in \mathbb{Z}$, which yields $f(rx) = rf(x)$ for any $r \in \mathbb{Q}$. Since f is continuous and every real number r can be approximated by a sequence of rational numbers, we find that $f(rx) = rf(x)$ for any $r \in \mathbb{R}$, and hence the claim.

Let us prove the uniqueness of such additive functions. Assume that there exists another additive function h satisfying

$$|f(x) - h(x)| \leq M, \quad \text{for all } x \in \mathbb{R}.$$

This implies that

$$|g(x) - h(x)| \leq M, \quad \text{for all } x \in \mathbb{R}.$$

However, an additive function is either zero or unbounded. Therefore, the additive function $g - h$ must vanish.

Comments 88 *Functions f satisfying the condition from the statement are called quasimorphisms.*

Notice that there exist additive functions $\varphi : \mathbb{R} \to \mathbb{R}$ that are not linear (not continuous of course). Examples can be constructed by choosing a basis of $\mathbb{R}$ as a vector space over $\mathbb{Q}$ and defining arbitrarily the values of the function on each vector of that basis.

Problem 4.39. Show that if f is differentiable and if

$$\lim_{t \to \infty} \left(f(t) + f'(t) \right) = 1,$$

then

$$\lim_{t \to \infty} f(t) = 1.$$

Solution 4.39. Let us show first that if $\lim_{t \to \infty}(f'(t) + \alpha f(t)) = 0$, where $a = \Re(\alpha) > 0$, then $\lim_{t \to \infty} f(t) = 0$. Let $\varepsilon > 0$ and $c < \infty$ be such that $|f'(t) + \alpha f(t)| \le a\varepsilon$ whenever $t \ge c$. Then

$$\left| \frac{\mathrm{d}}{\mathrm{d}t} \left(e^{\alpha t} f(t) \right) \right| = \left| e^{\alpha t} (f'(t) + \alpha f(t)) \right| \le a\varepsilon e^{at}$$

for $t \ge c$. Therefore, using the mean value theorem we have

$$|e^{\alpha t} f(t) - e^{\alpha c} f(c)| \le \varepsilon \left(e^{at} - e^{ac} \right), \quad \text{for } t > c.$$

This yields

$$|f(t)| \le e^{a(c-t)} \cdot |f(c)| + \varepsilon \left| 1 - e^{a(c-t)} \right|.$$

In particular, for sufficiently large t, we have $|f(t)| \le 2\varepsilon$, as claimed. Now the statement follows by applying this result to $f - 1$.

Comments 89 *Let $P(\mathcal{D})$ be a polynomial in the derivation operator $\mathcal{D} = \frac{\mathrm{d}}{\mathrm{d}t}$. Let us write*

$$P(\mathcal{D}) = (\mathcal{D} - \lambda_1)(\mathcal{D} - \lambda_2) \cdots (\mathcal{D} - \lambda_n).$$

The above result can be generalized as follows. Assume that $\Re\lambda_i < 0$, where $\Re$ denotes the real part; if the function f satisfies $\lim_{t \to \infty} P(\mathcal{D}) f(t) = 0$, then $\lim_{t \to \infty} f(t) = 0$. This is proved by recurrence on the degree of P. The recurrence step follows from the argument given above. In particular, the claim holds for $P(\mathcal{D}) = \mathcal{D}^2 + \mathcal{D} + 1$.

The condition $\Re\lambda_i < 0$ is necessary. In fact, we can consider $f(t) = e^{\lambda_i t}$, which satisfies $\lim_{t \to \infty} P(\mathcal{D}) f \to 0$. Also, one notices that $P(\mathcal{D}) = \mathcal{D}^0 + \mathcal{D}^1 + ... + \mathcal{D}^n$ does not fulfill this condition for $n \ge 3$.

Problem 4.40. Let c be a real number and let $f : \mathbb{R} \to \mathbb{R}$ be a smooth function of class $\mathcal{C}^3$ such that $\lim_{x \to \infty} f(x) = c$ and $\lim_{x \to \infty} f'''(x) = 0$. Show that $\lim_{n \to \infty} f'(x) = \lim_{x \to \infty} f''(x) = 0$.

Solution 4.40. According to Taylor's formula, there exist $\xi_x, \eta_x \in (0, 1)$ such that

$$f(x + 1) = f(x) + f'(x) + \frac{1}{2}f''(x) + \frac{1}{6}f'''(x + \xi_x),$$

$$f(x - 1) = f(x) - f'(x) + \frac{1}{2}f''(x) - \frac{1}{6}f'''(x + \eta_x).$$

Suitable linear combinations of these yield

$$f''(x) = f(x + 1) - 2f(x) + f(x - 1) - \frac{1}{6}f'''(x + \xi_x) + \frac{1}{6}f'''(x - \eta_x),$$

$$2f'(x) = f(x + 1) - f(x - 1) - \frac{1}{6}f'''(x + \xi_x) - \frac{1}{6}f'''(x - \eta_x).$$

Since $x + \xi_x \to \infty$ and $x - \eta_x \to \infty$ as $x \to \infty$, by passing to the limit, one obtains

$$\lim_{x \to \infty} f''(x) = c - 2c + c - \frac{1}{6} \cdot 0 + \frac{1}{6} \cdot 0 = 0,$$

$$\lim_{x \to \infty} f'(x) = \frac{1}{2}\left(c - c - \frac{1}{6} \cdot 0 - \frac{1}{6} \cdot 0\right) = 0.$$

Comments 90 *It can be proved that if $\lim_{x \to \infty} f(x)$ exists and if $f^{(n)}$ is bounded, then $\lim_{x \to \infty} f^{(k)}(x) = 0$ for any $1 \le k < n$. See for instance:*

- J. Littlewood: *The converse of Abel's theorem*, Proc. London Math. Soc. (2) 9 (1910–11), 434–448.

Problem 4.41. Prove that the following integral equation has at most one continuous solution on $[0, 1] \times [0, 1]$:

$$f(x, y) = 1 + \int_0^x \int_0^y f(u, v)\, du\, dv.$$

Solution 4.41. If f_1, f_2 are two solutions, then their difference g satisfies

$$g(x, y) = \int_0^x \int_0^y g(u, v)\, du\, dv.$$

Since g is continuous, its absolute value is bounded by some constant M on the square $[0, 1] \times [0, 1]$. Thus

$$|g(x, y)| \le \int_0^x \int_0^y M\, du\, dv = Mxy.$$

By induction, we show that $|g(x, y)| \le M\frac{x^n y^n}{(n!)^2}$. The induction step follows from

$$|g(x, y)| \le \int_0^x \int_0^y \frac{M u^n v^n}{(n!)^2}\, du\, dv = \frac{M(xy)^{n+1}}{((n+1)!)^2}.$$

If we fix x, y, then

$$0 \le |g(x, y)| \le \lim_{n \to} \frac{M x^n y^n}{n! n!} = 0,$$

and thus $g(x, y) \equiv 0$.

Comments 91 *There exists a continuous solution of this equation, given by*

$$f(x, y) = 1 + xy + \frac{x^2 y^2}{2!2!} + \frac{x^3 y^3}{3!3!} + \cdots = J_0\left(\sqrt{-4xy}\right) = J_0\left(2i\sqrt{xy}\right),$$

where J_0 is the Bessel function of order zero.

Problem 4.42. Find those $\lambda \in \mathbb{R}$ for which the functional equation

$$\int_0^1 \min(x, y) f(y) \, dy = \lambda f(x)$$

has a solution f that is nonzero and continuous on the interval $[0, 1]$. Find these solutions.

Solution 4.42. The equation can be written in the form

$$\lambda f(x) = \int_0^x y f(y) \, dy + x \int_x^1 f(y) \, dy.$$

If $\lambda \neq 0$, then f is differentiable, and after derivation, we obtain

$$\lambda f'(x) = x f(x) - x f(x) + \int_x^1 f(y) dy = \int_x^1 f(y) dy.$$

Thus f' is differentiable and

$$\lambda f''(x) = -f(x).$$

This implies that $f(x) = A \cos \mu x + B \sin \mu x$, where $\mu = \lambda^{-1/2}$ if $\lambda > 0$, and $f(x) = A \cosh \nu x + B \sinh \nu x$, where $\nu = (-\lambda)^{-1/2}$ if $\lambda < 0$. According to the hypothesis,

$$\lim_{n \to 0} f(x) = 0,$$

and hence $A = 0$ for any choice of λ. From the above, we also get

$$\lim_{n \to 0} f'(x) = 0,$$

and thus $B\mu \cos \mu = 0$ and $B\nu \cosh \nu = 0$. The last equation yields $B = 0$, and so we do not have nonzero solutions. For $\lambda > 0$, if $B \neq 0$, then $\cos \mu = 0$ and thus $\mu = (2k + 1)\frac{\pi}{2}$. Thus $\lambda = \mu^{-2} \frac{4}{(2k+1)^2 \pi^2}$, where $k \in \mathbb{Z}$, and

$$f(x) = B \cos(2k + 1)\frac{\pi}{2}x, \quad B \in \mathbb{R}.$$

If $\lambda = 0$, the same method gives us $f(x) = 0$.

8 Analysis Solutions   211

Comments 92 *We proved that the integral operator $T : C^0(0, 1) \to C^0(0, 1)$ defined on the space $C^0(0, 1)$ of continuous functions on $(0, 1)$ and given by the formula*

$$(Tf)(x) = \int_0^1 \min(x, y) f(y) dy$$

has the eigenvalues $\frac{4}{(2k+1)^2\pi^2}$ and the eigenvectors $\cos(2k + 1)\frac{\pi}{2}x$.

Problem 4.43. Let X be an unbounded subset of the real numbers $\mathbb{R}$. Prove that the set

$$A_X = \{t \in \mathbb{R}; tX \text{ is dense modulo } 1\}$$

is dense in $\mathbb{R}$.

Solution 4.43. Let $\{O_n\}_{n\in\mathbb{Z}_+}$ be a countable base of $(0, 1)$, $O_n \neq \emptyset$. Let p be the canonical projection $p : \mathbb{R} \to \mathbb{R}/\mathbb{Z}$. Define

$$A_n = \{r \in \mathbb{R}; p(rX) \cap O_n \neq \emptyset\}.$$

This amounts to

$$A_n = \bigcup_{x \in X - \{0\}} \frac{1}{x} p^{-1}(\{O_n\}),$$

and thus A_n is open. If F is an interval of length $f > 0$, then there exists $x \in X$ such that $xf > 1$, which implies that $p(xF) = \mathbb{R}/\mathbb{Z}$ and thus there exists $r \in F$ with $p(rx) \in O_n$. This proves that A_n is dense in $\mathbb{R}$. Since

$$A_X = \bigcap_{n=0}^{\infty} A_n,$$

we obtain that A_X is also dense in $\mathbb{R}$, as the intersection of open dense subsets.

Problem 4.44. Consider $P(z) = z^n + a_1 z^{n-1} + \cdots + a_n$, where $a_i \in \mathbb{C}$. If $|P(z)| = 1$ for all z satisfying $|z| = 1$, then $a_1 = \cdots = a_n = 0$.

Solution 4.44. Consider the polynomial $Q(z) = 1 + a_1 z + \cdots + a_n z^n$, which can be written as $Q(z) = P(\bar{z})$, when $|z| = 1$. This implies that $|Q(z)| = 1$ for all z of unit modulus. Moreover, $Q(0) = 1$, and so Q has a value in the interior of the disk whose modulus is equal to its maximum on the boundary circle. According to the maximum principle for holomorphic functions, Q has to be constant, $Q(z) = 1$, whence the claim.

Comments 93 *W. Blaschke has studied the complex analytic (i.e., holomorphic) functions $f : D \to \mathbb{C}$ on the unit disk D that are continuous on D and have $|f(z)| = 1$ for all z satisfying $|z| = 1$. He has shown that such a function has the form*

$$f = \sigma \prod_{k=1}^{n} \frac{z - b_k}{1 - \bar{b}_k z},$$

where σ is a constant of modulus one and $b_k \in \mathbb{C}$ are such that $|b_k| < 1$.

Problem 4.45. Let $I \subset \mathbb{R}$ be an interval and $u, v : I \to \mathbb{R}$ smooth functions satisfying the equations

$$u''(x) + A(x)u(x) = 0, \quad v''(x) + B(x)v(x) = 0,$$

where A, B are continuous on I and $A(x) \geq B(x)$ for all $x \in I$. Assume that v is not identically zero. If $\alpha < \beta$ are roots of v, then there exists a root of u that lies within the interval (α, β), unless $A(x) = B(x)$, in which case u and v are proportional for $\alpha \leq x \leq \beta$.

Solution 4.45. The roots of v are isolated and we can suppose that $v|(\alpha, \beta) > 0$. Assume that $u|(\alpha, \beta)$, so that after a possible sign change, $u|(\alpha, \beta) > 0$. Also, we have $v'(\alpha) \geq 0$, $v(\beta) \leq 0$, and since v is not identically zero, then v and v' do not have common roots; thus $v'(\alpha) > 0$ and $v'(\beta) < 0$. We consider

$$w(x) = u(x)v'(x) - u'(x)v(x).$$

We have

$$w'(x) = u(x)v''(x) - u''(x)v(x) = (A(x) - B(x))uv \geq 0,$$

and therefore w is increasing; so $w(\alpha) \leq w(\beta)$ and thus $u(\alpha)v'(\alpha) \leq u(\beta)v'(\beta)$. Since $u(\alpha) \geq 0$, $u(\beta) \geq 0$, $v'(\alpha) > 0$, $v'(\beta) < 0$, we obtain $u(\alpha) = u(\beta) = 0$, and so $w(\alpha) = w(\beta) = 0$. Now w is increasing and hence $w \equiv 0$ and thus $w' \equiv 0$, which yields $A(x) = B(x)$. Moreover, for $\alpha < x < \beta$ we have $\frac{w(x)}{v^2(x)} = (\frac{u}{v})'$, and hence $\frac{u}{v}$ is constant.

Comments 94 *This result is known as Sturm's comparison theorem in differential equations.*

Problem 4.46. Let V be a finite-dimensional real vector space and $f : V \to \mathbb{R}$ a continuous mapping. For any basis $B = \{b_1, b_2, \ldots, b_n\}$ of V, consider the set

$$E_B = \{z_1 b_1 + \cdots + z_n b_n, \text{ where } z_i \in \mathbb{Z}\}.$$

Show that if f is bounded on E_B for any choice of the basis B, then f is bounded on V.

Solution 4.46. We solve first the case $n = 1$. Given any pair of positive numbers $0 < a < b$, there exists some $r = r(a, b)$ (depending on a, b) such that for any $x > r$, one can find $m \in \mathbb{Z}$, with $ma \leq x \leq mb$. In fact, the intervals (ma, mb), for integral m, will cover all of $\mathbb{R}$ but a compact set. Thus, for any $x > r$, one can find an interval (c, d) containing x, and some $m \in \mathbb{Z}$ such that $a < \frac{c}{m} < \frac{d}{m} < b$.

Assume that f is unbounded. Let us consider an interval $I_1 = [a_1, b_1]$ with the property that $f(x) > 1$, for $x \in I_1$. Since f is unbounded, there exist $x_1 > r(a_1, b_1)$ arbitrarily large such that $f(x_1) > 4$ and $m_1 \in \mathbb{Z}$ such that $m_1 a_1 < x_1 < m_1 b_1$. Moreover, f is continuous and so $f(x) > 2$ for any $x \in [c_1, d_1]$, for some small interval around x_1, which can be chosen to be contained in $(m_1 a_1, m_1 b_1)$. Set $I_2 =$

$\{x \in I_1; m_1 x \in [c_1, d_1]\}$. Thus $I_2 \subset I_1$ is a compact nonempty interval with the property that $f(m_1 x) > 2$ for any $x \in I_2$.

We define in this way a nested sequence of compact intervals $I_1 \subset I_2 \subset I_3 \subset \cdots I_k \subset \cdots$, and a sequence of integers $m_1 < m_2 < m_3 < \cdots$ with the property that $f(x) > 2^k$ for $x \in m_k I_k$. This nested sequence of nontrivial compact intervals must have nontrivial intersection. Let $a \in \cap_{k=1}^{\infty} I_k$. Then the restriction of f to $a\mathbb{Z}$ is unbounded, since for any k, there exists some m_k with $f(m_k a) > 2^k$, by making use of the fact that $a \in I_k$.

Let now deal with the general case. Let $z_k \in V$ be a sequence for which $f(z_k)$ is unbounded, e.g., $f(z_k) > 2^{k+1}$. Then z_k is unbounded, because f is continuous.

Let $B = \{b_1, \ldots, b_n\}$ be a basis of B and write $z_k = \sum_{i=1}^{n} c_{ik} b_i$ in the basis B. There exists at least one $i_0 \in \{1, 2, \ldots, n\}$ such that the sequence $c_{i_0 k}$ is unbounded; otherwise, z_k would be bounded. One can slightly modify the basis B so that this condition is satisfied for all i. In fact, assume that c_{ik} is unbounded for $1 \le i \le m < n$ and c_{ik} is bounded for $m + 1 \le i \le n$. Take then the new basis vectors

$$b_1' = b_1 - \sum_{i=m+1}^{n} b_i, \quad b_i' = b_i, \text{ when } i \ge 2.$$

In the new basis, one can express the vector z_k as

$$z_k = \sum_{i=1}^{m} c_{ik} b_i' + \sum_{i=m+1}^{n} (c_{ik} + c_{1k}) b_i',$$

and all components are now unbounded.

We consider now the intervals J_{ik} with the property that $U_k = \{x; x = \sum_{i=1}^{n} \alpha_{ik} b_i,$ with $\alpha_{ik} \in J_{ik}\}$ is a neighborhood of z_k and $f(x) > 2^k$, for all $x \in U_k$.

The argument for $n = 1$ shows that there exists a sequence of integers (m_{ik}) going to infinity such that we have a nested sequence of nontrivial compact intervals

$$\cdots \subset \frac{1}{m_{ik+1}} J_{ik+1} \subset \frac{1}{m_{ik}} J_{ik} \subset \cdots \subset J_{i1}.$$

Take

$$\gamma_i \in \bigcap_{k=1}^{\infty} \frac{1}{m_{ik}} J_{ik}$$

and construct a new basis $\widetilde{B} = \{\gamma_i b_i, i \in \{1, 2, \ldots, n\}\}$. It follows that f is unbounded on $E_{\widetilde{B}}$.

Problem 4.47. It is known that if $f, g : \mathbb{C} \to \mathbb{C}$ are entire functions without common zeros then there exist entire functions $a, b : \mathbb{C} \to \mathbb{C}$ such that $a(z) f(z) + b(z) g(z) = 1$ for all $z \in \mathbb{C}$.

1. Prove that we can choose $a(z)$ without any zeros.
2. Is it possible to choose both a and b without zeros?

Solution 4.47. 1. We know that $Af + Bg = 1$ for some entire functions A and B. Therefore, for any entire function λ, we have also

$$(A + \lambda g)f + (B - \lambda f)g = 1.$$

We now choose $a = A + \lambda g$ such that $A + \lambda g = e^h$. If z_0 is a zero of g, we need to have $e^{h(z_0)} = A(z_0)$ (with the same multiplicity). This is possible from the interpolation theorem for entire functions, because $A(z_0)f(z_0) = 1$.

2. Let $f, g \in \mathbb{C}[z]$ be nonconstant polynomials of different degrees. We will show that there are no entire functions a, b such that $af + bg = 1$ and a, b have no zeros. There are no integer nonconstant functions.

First, there exist nonzero polynomials A and B such that $Af + Bg = 1$. As above, for any entire function $\lambda : \mathbb{C} \to \mathbb{C}$, one can construct other solutions, namely $a = A - \lambda g$ and $b = B + \lambda f$. Let us now show that any pair a, b can be obtained in this way. We have $(A - a)f + (B - b)g = 0$. Thus the set of zeros of $(A - a)$ contains the set of zeros of g (since f and g have no common zeros). In particular, there exists an entire function λ such that $A - a = \lambda g$. In a similar way, there exists an entire function μ such that $B - b = \mu f$. The condition above shows that $\lambda = \mu$, as claimed.

If a, b do not have zeros, then the meromorphic function $F = \frac{1}{\lambda}$ must be distinct everywhere from the rational functions a_1, a_2, a_3 given below:

$$a_1(z) = 0, \quad a_2(z) = g(z)/A(z), \quad a_3(z) = -f(z)/B(z).$$

Consider the meromorphic function

$$G = \frac{F - a_1}{F - a_2} \cdot \frac{a_3 - a_2}{a_3 - a_1}.$$

The points where $G(z) \in \{0, 1, \infty\}$ are among those satisfying one of the conditions $a_1(z) = a_2(z)$, $a_1(z) = a_3(z)$, $a_2(z) = a_3(z)$, and therefore they are finitely many. According to Picard's theorem, G does not have an essential singularity at ∞ and thus G is rational. This implies that F is rational; hence λf is a polynomial and thus a and b are polynomials. But polynomials without zeros are constant, and this is impossible since f and g have different degrees.

Problem 4.48. Consider a compact set $X \subset \mathbb{R}$. Show that a necessary and sufficient condition for the existence of a monic nonconstant polynomial with real coefficients $h \in \mathbb{R}[x]$ such that $|h(x)| < 1$ for all $x \in X$ is the existence of monic nonconstant polynomial $g(x) \in \mathbb{R}[x]$ such that $|g(x)| < 2$ for all $x \in X$. Prove that 2 is the maximal number with this property.

Solution 4.48. Let $\mathcal{P}(X)$ be the Banach space of real polynomials endowed with the norm $\|\theta\| = \sup_{x \in X} |\theta(x)|$. We define the nonlinear operator

$$A : \mathcal{P}(x) \to \mathcal{P}(x)$$

given by

$$(Af(x)) = f^2(x) - \frac{1}{2}\|f\|^2, \text{ for } f \in \mathcal{P}(X).$$

If f is monic, then Af is also monic and we have $\|Af\| = \frac{1}{2}\|f\|^2$. An easy induction implies that

$$\|A^n f\| = \frac{2\|f\|^{2n}}{2^{2n}}.$$

Moreover, if $\|f\| < 2$, then there exists some n for which $\|A^n f\| < 1$. This proves the first part.

Let $X = [-2, 2]$. The norm of $p(x) = x$ on X is $\|p\| = 2$. We will prove that any monic polynomial on X has norm at least 2. Assume the contrary, namely that there exists $f \in \mathcal{P}(x)$, monic, with $\|f\| < 2$. We have two intermediate results:

1. There exists an operator $T : \mathcal{P}(X) \to \mathcal{P}(X)$ that takes monic polynomials of degree $2k$ to monic polynomials of degree k such that

$$\|TP\| < \|P\|.$$

2. Assuming the existence of a monic polynomial of norm $\|f\| < 2$, there exists $n \in \mathbb{Z}_+$ and a monic polynomial g, of degree 2^n, with $\|g\| < 2$.

By points 1 and 2 above there exists a monic polynomial g of degree 2^n with $\|g\| < 2$ on $[-2, 2]$, which implies that there exists a monic polynomial of degree 1 such that $\|g\| < 2$ on $[-2, 2]$, which is obviously false.

Proof of claim 1. If $h \in \mathcal{P}(x)$, let us consider $\tilde{h}(x) = \frac{1}{2}(h(x) + h(-x))$, which is even and thus can be written as a polynomial in x^2. Thus $\tilde{h}(x) = Q(x^2)$, where now $Q : [0, 4] \to \mathbb{R}$. Define $T(h)(x) = Q(x + 2)$. It is clear that T has the desired properties.

Proof of claim 2. If f has degree $2^m p$, where p is odd, then $t = T^m f$ is monic of degree p, $\|t\| < 2$. There exists $n \in \mathbb{Z}_+$ such that p divides $2^n - 1$, and so $2^n - 1 = pq$. We define $g \in \mathcal{P}(x)$ by means of $g = t(x)^q x$. It is immediate that $\|g\| < 2$, and the proof is complete.

9

Glossary

9.1 Compendium of Triangle Basic Terminology and Formulas

9.1.1 Lengths in a Triangle

We used the standard notation for the important features of a triangle, as follows:

- a, b, c the sides lengths
- A, B, C the angles
- m_a, m_b, m_c the medians
- h_a, h_b, h_c the altitudes
- w_a, w_b, w_c the angle bisectors
- R the radius of the circumcircle
- r the radius of the incircle
- r_a, r_b, r_c the radii of the extrinsic circles
- S the area
- p the semiperimeter

We collect below some useful identities between these quantities:

- $S = \frac{1}{2}bc \sin A = (p(p-a)(p-b)(p-c))^{1/2} = 4Rr \cos \frac{A}{2} \cos \frac{B}{2} \cos \frac{C}{2}$,
- $S = rp$,
- $4RS = abc$,
- $r_a(p-a) = rp = S, \quad r_a = p \tan \frac{A}{2}$,
- $4R + r = r_a + r_b + r_c$,
- $ah_a = 2S$,
- $h_a h_b h_c = \frac{8S^3}{abc} = \frac{2S^2}{R}$,
- $\frac{1}{h_a} + \frac{1}{h_b} + \frac{1}{h_c} = \frac{1}{r_a} + \frac{1}{r_b} + \frac{1}{r_c} = \frac{1}{r}$,
- $m_a^2 + m_b^2 + m_c^2 = \frac{3}{4}(a^2 + b^2 + c^2)$,
- $w_a = 2\frac{\sqrt{bc}}{b+c} \sqrt{p(p-a)}$,
- $r_a = \sqrt{\frac{p(p-b)(p-c)}{p-a}}$,

- $a^2 + b^2 + c^2 = 8R^2(1 + \cos A \cos B \cos C)$,
- $\sin \frac{A}{2} = \sqrt{\frac{(p-b)(p-c)}{bc}}$,
- $\cos A + \cos B + \cos C = 1 + 4 \sin \frac{A}{2} \sin \frac{B}{2} \sin \frac{C}{2} = 1 + \frac{r}{R}$,
- $\tan \frac{A}{2} = \sqrt{\frac{(p-b)(p-c)}{p(p-a)}}$,
- $R = \frac{a}{2 \sin A} = \frac{b}{2 \sin B} = \frac{c}{2 \sin C}$,
- $4m_a^2 = 2b^2 + 2c^2 - a^2$.

9.1.2 Important Points and Lines in a Triangle

- The *circumcenter O* is the center of the triangle's circumcircle. It can be found as the intersection of the perpendicular bisectors of the sides.
- The *incenter I* is the center of the incircle and thus the intersection of the angle bisectors.
- The *orthocenter H* is the intersection of the three altitudes of the triangle.
- The *centroid G* is the intersection of the three medians.
- The *excenter J_A* is the center of the excircle lying in the angle $\widehat{BAC}$, outside the triangle ABC and tangent to the lines AB, AC and to the segment BC. There are three excenters, J_A, J_B, J_C, using similar definitions.
- If M_A, M_B, M_C are the midpoints of BC, CA, AB respectively, then the three lines $M_A J_A, M_B J_B, M_C J_C$ are concurrent, and their intersection point is called the *mittenpunkt*.
- If S_A, S_B, S_C are the contact points of the excircles with BC, CA, AB respectively, then AS_A, BS_B, CS_C are concurrent and their intersection is called the *Nagel point Na*.
- Let T_A, T_B, T_C be the contact points of the incircle with sides BC, CA, AB respectively. Then the lines AT_A, BT_B, CT_C are concurrent, and their intersection point is called the *Gergonne point Ge*.
- The *isogonal conjugate* of the point X is the intersection of the three concurrent lines that are the reflections of AX, BX, CX about the angle bisectors of A, B, C respectively.
- The intersection point of the symmedians is called the *symmedian point* (or the *Lemoine point*). It is the isogonal conjugate of the centroid G.
- Two lines passing through the vertex A of a triangle ABC are called *isotomic* if they cut the side BC in points symmetric with respect to the midpoint of BC. Given a point P, the isotomic lines of the Cevians AP, BP, and CP meet at a point P' called the *isotomic point* of P.
- The *Fermat point F* of an acute triangle is the point that minimizes the sum of the distances to the vertices. If we draw equilateral triangles BCA', ABC', ACB' on the outside of the triangle ABC, then the three lines AA', BB', CC' are concurrent at F.
- The *Euler line*: the points O, G, H are collinear and $|HG| : |GO| = 2 : 1$.

- The Gergonne point Ge, the triangle centroid G, and the mittenpunkt M are collinear, with $|Ge\,G| : |GM| = 2 : 1$.
- The Nagel line: the points I, G, Na are collinear and $|Na\,G| : |GI| = 2 : 1$.

Kimberling tabulated and enumerated properties of more than 1477 centers in a triangle (later Brisse extended the list to 2001 items). The interested reader might consult his updated web page:

- C. Kimberling: see
  http://faculty.evansville.edu/ck6/encyclopedia/ETC.html

9.1.3 Coordinates in a Triangle

We are given a reference triangle ABC.

Barycentric coordinates. The barycentric coordinates of the point P are given by an ordered triple $F_P = (t_A, t_B, t_C)$ (up to nonzero scalar multiplication) such that P is the centroid of the system consisting of the three masses t_A, t_B, t_C located at the respective vertices A, B, C of the triangle, i.e., the mass t_A is located at A, etc. These coordinates were introduced by Möbius in 1827. The coordinates t_A are proportional to the areas area(PBC). If they are actually equal to these areas, they are called *homogeneous barycentric coordinates*.

Here are the coordinates of the principal points in a triangle.

- $F_G = (1, 1, 1)$
- $F_O = \left(a^2(b^2 + c^2 - a^2), b^2(a^2 + c^2 - b^2), c^2(a^2 + b^2 - c^2)\right)$
- $F_{J_A} = (-a, b, c)$, $F_I = (a, b, c)$
- $F_{Ge} = ((p - b)(p - c), (p - c)(p - a), (p - a)(p - b))$
- $F_{Na} = (p - a, p - b, p - c)$
- $F_H = \left((a^2 + c^2 - b^2)(a^2 + b^2 - c^2), (a^2 + b^2 - c^2)(b^2 + c^2 - a^2), (a^2 + c^2 - b^2)(b^2 + c^2 - a^2)\right)$
- $F_K = (a^2, b^2, c^2)$

The line determined by the points (s_1, s_2, s_2) and (t_1, t_2, t_3) has the equation

$$\det \begin{pmatrix} s_1 & s_2 & s_3 \\ t_1 & t_2 & t_3 \\ x_1 & x_2 & x_3 \end{pmatrix} = 0.$$

Trilinear coordinates. The trilinear coordinates of a point P are given by an ordered triple of numbers proportional to the directed distance from P to the sides, up to multiplication by a nonzero scalar. If we normalize them so that they give the distances to the sides, then these are called the *exact trilinear coordinates*. Trilinear coordinates are also known as *homogeneous coordinates*. They were introduced by Plücker in 1835. For instance, the vertices have trilinear coordinates $(1 : 0 : 0)$, $(0 : 1 : 0)$, and $(0 : 0 : 1)$.

For most important points in a triangle, the trilinear coordinates are given by triples of the form $(f(a, b, c) : f(b, c, a) : f(c, a, b))$, for some function f on the sides a, b, c. For instance, we have $f_H(a, b, c) = \cos B \cos C$, $f_G(a, b, c) = \frac{1}{a}$, $f_O(a, b, c) = \cos A$, $f_I(a, b, c) = 1$. For the Fermat point, we can take $f_F(a, b, c) = \sin(A + \frac{\pi}{3})$.

The homogeneous barycentric coordinates corresponding to $(x : y : z)$ are (ax, by, cz). The trilinear coordinates of the isogonal conjugate of $(x : y : z)$ are $\left(\frac{1}{x}, \frac{1}{y}, \frac{1}{z}\right)$.

9.2 Appendix: Pell's Equation

Pell's equation is the Diophantine equation

$$x^2 - dy^2 = 1.$$

The first treatment of this equation was actually given by Lord William Brouncker in 1657, but his solution was erroneously ascribed to John Pell by Euler. Lagrange developed the theory of continued fractions, which gives the actual method of finding the minimal solution, and published the first proof in 1766. However, some methods (the cyclic method) were already known to the Indian mathematicians Brahmagupta and Bhaskara in the twelfth century.

General solutions. Pell's equation has infinitely many solutions if d is not a perfect square and none otherwise. If (x_0, y_0) denotes its minimal solution (called also the fundamental solution) different from $(1, 0)$, then all solutions (x_n, y_n) are obtained from it by means of the recurrence relations

$$x_{n+1} = x_0 x_n + d y_0 y_n, \quad y_{n+1} = y_0 x_n + x_0 y_n, \quad x_1 = x_0, \; y_1 = y_0.$$

The fundamental solution. Thus the main point is finding the fundamental solution. This can be achieved using continued fractions. According to Lagrange, the continued fraction of a quadratic irrational number (i.e., a real number satisfying a quadratic equation with rational coefficients that is not rational itself) is periodic after some point; in our case, even more can be obtained, namely

$$\sqrt{d} = \left[a_0, \overline{a_1, \ldots, a_m, 2a_0}\right].$$

The numbers a_j can be calculated recursively, as follows. Set $a_0 = \left[\sqrt{d}\right]$ and construct the sequences

$$
\begin{aligned}
P_0 &= 0, & Q_0 &= 1, \\
P_1 &= a_0, & Q_1 &= d - a_0^2, \\
P_n &= a_{n-1} Q_{n-1} - P_{n-1}, & Q_n &= \frac{d - P_n^2}{Q_{n-1}}.
\end{aligned}
$$

Then

$$a_n = \left[\frac{a_0 + P_n}{Q_n}\right].$$

Furthermore, consider the sequences

$$p_0 = a_0, \qquad\qquad q_0 = 1,$$
$$p_1 = a_0a_1 + 1, \qquad\qquad q_1 = a_1,$$
$$p_n = a_n p_{n-1} + p_{n-2}, \quad q_n = a_n q_{n-1} + q_{n-2}.$$

Then we have the identities

$$p_n^2 - dq_n^2 = (-1)^{n+1} Q_{n+1},$$

which permit one to construct solutions to the Pell-type equations.

If $a_{m+1} = 2a_0$, as happens for a quadratic irrational, then $Q_{m+1} = 1$. In particular, we find that the fundamental solution is determined by the parity of m, as follows:

$$x_0 = p_m, \quad y_0 = q_m, \quad \text{if } m \text{ is odd},$$

$$x_0 = p_{2m+1}, \quad y_0 = q_{2m+1}, \quad \text{if } m \text{ is even}.$$

Observe that $p_r/q_r = [a_0, a_1, \ldots, a_r]$. The general solution (x_n, y_n) is obtained by means of the trick

$$x^2 - dy_2 = (x_0^2 - dy_0^2)^n = 1,$$

and setting

$$x + \sqrt{d}\,y = \left(x_0 + \sqrt{d}\,y_0\right)^n, \quad x - \sqrt{d}\,y = \left(x_0 - \sqrt{d}\,y_0\right)^n,$$

we obtain the family of solutions

$$x_n = \frac{\left(x_0 + \sqrt{d}\,y_0\right)^n + \left(x_0 - \sqrt{d}\,y_0\right)^n}{2}, \quad y_n = \frac{\left(x_0 + \sqrt{d}\,y_0\right)^n - \left(x_0 - \sqrt{d}\,y_0\right)^n}{2}.$$

The negative Pell equation. The Pell-like equation

$$x^2 - dy^2 = -1$$

can also be solved, but it does not always have solutions. When it has one, then it has infinitely many. A necessary condition for this equation to be solvable is that all odd prime factors of d be of the form $4k + 1$, and that $d \not\equiv 0 \pmod 4$. However, these conditions are not sufficient, as can be seen from the equation for $d = 34$, which has no solutions. The method of continued fractions works for these equations as well. In fact, for this Pell-type equation, we have the fundamental solution

$$x_0 = p_m, \quad y_0 = q_m, \quad \text{if } m \text{ is even}.$$

The equation does not have solutions if m is odd. Furthermore, the general solution is given again by

$$x_n = \frac{\left(x_0 + \sqrt{d}\,y_0\right)^n + \left(x_0 - \sqrt{d}\,y_0\right)^n}{2}, \quad y_n = \frac{\left(x_0 + \sqrt{d}\,y_0\right)^n - \left(x_0 - \sqrt{d}\,y_0\right)^n}{2},$$

but *only* for odd n.

Pell-like equations. The Pell-like equation

$$x^2 - dy^2 = c$$

can be solved using the same ideas. If $|c| < \sqrt{d}$, then the equation has solutions if and only if

$$c \in \left\{ Q_0, -Q_1, \ldots, (-1)^{m+1} Q_{m+1} \right\}.$$

Moreover, if $c > \sqrt{d}$, then the procedure is significantly more complicated. See for instance the comments of Dickson on this subject. It is clear that using the solutions (x_n, y_n) to the companion Pell equation $x^2 - dy^2 = 1$ and a particular solution (u, v) of the general Pell-type equation from above, we are able to find infinitely many solutions by means of the recurrence

$$u_n = x_n u \pm d y_n v, \quad v_n = x_n v \pm y_n u.$$

However, even if we start with the minimal solution, we cannot always obtain all solutions of the equation using this recurrence. The reason is that we might well have several fundamental solutions. For instance, if $d = 10$ and $c = 9$, then we have the fundamental solutions $(7, 2)$, $(13, 4)$ and $(57, 18)$.

The general equation

$$ax^2 - by^2 = c$$

can be reduced to the former equation by setting $d = ab$, $x' = ax$, $c' = ac$ and looking for those solutions where x' is divisible by a.

The degree-two equation in two unknowns. Moreover, the general degree-two equation

$$ax^2 + bxy + cy^2 + dx + cy + f = 0$$

where $a, b, c, d, e, f \in \mathbb{Z}$, can be reduced to either Pell-type equations or elliptic equations.

1. The case $\Delta = b^2 - 4ac < 0$. The equation represents an ellipse in the plane, and therefore it has only finitely many integer solutions. These equations were solved by Gauss and earlier considered by Euler. According to Dickson one can find all its solutions using the solutions to Pell's equations.

2. If $\Delta = 0$, the parabolic-type equation becomes

$$\frac{1}{4a} \left((2ax + by + d)^2 + 2(ae - bd)y + 4af - d^2 \right) = 0.$$

If $2ae - bd = 0$, then the equation is equivalent, through a linear transformation, to

$$(2ax + by + d)^2 = d^2 - 4af;$$

hence the parabola degenerates into two lines. If $d^2 - 4af$ is a perfect square, then we have two infinite families of solutions $2ax + by + d = \pm\sqrt{d^2 - 4af}$; otherwise, there are no integer solutions.

If $2ae - bd \neq 0$, then we have to solve the equation $2(ae - bd)y = -t^2 + d^2 - 4af$, for an arbitrary parameter $t \in \mathbb{Z}$. This depends only on computing the residues modulo $2(ae - bd)$ of the perfect squares, and yields infinitely many solutions if one exists. Furthermore, $2ax = t - d - by$ determines x.

3. If $\Delta > 0$, then the hyperbolic-type equation becomes:

$$a(\Delta x - 2cd + bc)^2 + b(\Delta x - 2cd + bc)(\Delta y + 2ac + bd) + c(\Delta y - 2ac + bd)^2$$
$$= -\Delta(ac^2 + cd^2 + ef^2 - bdc - 4acf),$$

which, through a transformation $u = \Delta x - 2cd + bc, v = \Delta y - 2ac + bd$, becomes

$$au^2 + buv + cv^2 = h.$$

This equation can be reduced to a Pell equation. It will have then either zero or infinitely many solutions.

The Ankeny–Artin–Chowla Conjecture. Although the Pell equation seems to be well understood, there are many subtle questions concerning its solutions. Here is a problem that has resisted to all efforts until now.

1. If p is a prime $p \equiv 1 \pmod 4$ and y_0 is the smallest positive value of y such that

$$x^2 - py^2 = -4,$$

then $y_0 \not\equiv 0 \pmod p$.

2. If p is a prime $p \equiv 3 \pmod 4$ and y_0 is the smallest positive value of y such that

$$x^2 - py^2 = 1,$$

then $y_0 \not\equiv 0 \pmod p$.

The first conjecture is due to Ankeny, Artin, and Chowla, and the second to Mordell.

The Thue theorem for higher degree. The situation of equations of higher degree is completely different. Let

$$f(x, y) = a_0 x^n + a_1 x^{n-1}y + a_2 x^{n-2}y^2 + \cdots + a_n y^n$$

be a homogeneous form of degree n. If $n \geq 3$ and f is irreducible over $\mathbb{Q}$, then the equation

$$f(x, y) = g(x, y),$$

where $g(x, y)$ is an arbitrary form of degree $n - 2$ (not necessarily homogeneous), has only finitely many solutions. This is a celebrated theorem due to Thue and based on results of Siegel and Roth. Later A. Baker gave explicit bounds on the size of the solutions for a wide class of such f.

- L.E. Dickson: *Pell equation: $ax^2 + by^2 + c$ made square*, Chapter 12 in *History of the Theory of Numbers*, Vol. 2: Diophantine Analysis, New York, Chelsea, 341–400, 1952.
- H.W. Lenstra, Jr.: *Solving the Pell equation*, Notices Amer. Math. Soc. 49 (2002), 2, 182–192.

Index of Mathematical Results

Index of Mathematical Terms

Index of Topics and Methods

Classical arithmetic functions and applications: 1.4, 1.29, 1.34, 1.36, 1.42, 1.43, 1.44, 1.45, 1.46, 1.69

Congruences and divisibility: 1.1, 1.3, 1.8, 1.11, 1.15, 1.17, 1.22, 1.24, 1.27, 1.38, 1.48, 1.50, 1.51, 1.67, 1.68, 2.7, 2.45

Rational and algebraic numbers: 1.47, 2.14, 2.15

Polynomials: 1.2, 1.7, 1.30, 2.2, 2.4, 2.9, 2.12, 2.13, 2.24

Estimates of arithmetically defined functions: 1.5, 1.6, 1.23, 1.28, 1.37, 1.55, 1.65, 1.66, 2.25, 2.35, 2.47

Symmetric functions: 2.1, 2.8

General Diophantine equations (Pythagorean, exponential): 1.12, 1.14, 1.16, 1.20, 1.25, 1.33, 1.35, 1.39, 1.41, 1.49, 1.52, 1.56, 1.57, 1.60

Pell's equation and its applications: 1.13, 1.58, 1.59

Continued fractions: 1.13

Prime numbers, Chebyshev theorem: 1.9, 1.18, 1.19, 1.21, 1.32, 1.54, 2.46

Arithmetic progressions: 1.10, 1.26, 1.40, 2.5, 2.21

Representing integers by algebraic functions, Waring problem: 1.31, 1.53, 1.61, 1.62, 1.63, 1.64

Algebraic number theory: 1.70

Groups: 2.10

Combinatorial identities: 2.33, 2.34

Algebraic equations: 2.3, 2.16

Linear algebra, matrices, determinants: 2.6, 2.11, 2.17, 2.18, 2.40, 2.43

Combinatorial properties of sets of integers: 2.26, 2.27, 2.28, 2.29, 2.41, 2.44

Sequences of integers: 2.19, 2.42, 2.52, 2.53

Counting problems: 2.23, 3.21

Extremal problems concerning sets of integers: 2.36, 2.37, 2.48, 2.49, 2.50, 2.51

Printed In The United States Of America